W0107016

Editors

Lecture Notes in Chemistry

Edited by G. Berthier M. J. S. Dewar H. Fischer
K. Fukui G. G. Hall H. Hartmann H. H. Jaffé J. Jortner
W. Kutzelnigg K. Ruedenberg E. Scrocco

29

N.D. Epiotis

With J.R. Larson and H.L. Eaton

Unified Valence Bond Theory of Electronic Structure

Springer-Verlag
Berlin Heidelberg New York 1982

Authors
N.D. Epiotis
Department of Chemistry, University of Washington
Seattle, Washington 98195, USA

J.R. Larson
Department of Chemistry, University of Chicago
Chicago, Illinois 60637, USA

H.L. Eaton
Department of Chemistry, University of Washington
Seattle, Washington 98195, USA

ISBN-13:978-3-540-11491-8 e-ISBN-13:978-3-642-93213-7
DOI: 10.1007/978-3-642-93213-7

2152/3140-543210

FOREWORD

In the last fifty years, computational chemistry has made impressive strides. Hückel MO computations were rapidly succeeded by semiempirical monodeterminantal Self Consistent Field (SCF) MO calculations which now give way to high quality *ab initio* calculations of the poly-determinantal SCF-MO and Generalized VB variety. By contrast, no analogous progress has been made in the area of the qualitative theory of chemical bonding. In fact, more than a half-century after the exposition of Hückel MO theory the conceptual superstructure of chemistry is still founded on it. This is made glaringly evident by the fact that highly sophisticated computations are still interpreted with primitive Hückel MO theory, despite the fact that most chemists are well aware of its formal deficiencies. The current popularity of qualitative MO theory among experimentalists is not the result of formal advances but rather the consequence of stimulating application of old MO theoretical concepts. This work attemps to improve this situation by outlining a qualitative theory of chemical bonding which operates at a high level of theoretical sophistication. It was first presented at the NATO Advanced Study Institute on "Topics in Theoretical Organic Chemistry" in Gargnano, Italy, in June 1979, and in other international meetings and conferences, colloquia, and informal gatherings in the period of time following the Gargnano meeting. It was also presented in a seminar given at the University of Washington in October 1980. It can be said that this work is the result of the natural evolution of the type of thinking introduced many years ago in an article published in Angewandte Chemie and then pursued further in a series of papers and a monograph entitled "Theory of Organic Reactions" published by Springer-Verlag. It represents our total abandonment of MO theory as an interpretative and predictive tool and a call for a shift to sound, as opposed to intuitive, VB theory.

In developing a new conceptual approach one ultimately has to come to grips with presentation, so to speak, problems. In submitting this work to public scrutiny, we recognize two such problems. The first one has to do with the fact that the theory we are attempting to popularize is essentially a VB-type theory and, in an age of MO theory dominance, VB theoretical principles are hardly familiar to most chemists. We have tried to counteract this problem by being as explicit as possible in developing the theory while trying to simplify things at the same time. The second problem has to do with the understandable skepticism with which new ideas are met. The reader may legitimately ask: Is it worth the time mastering a new "language" when there is no guarantee that the allegedly new approach accomplishes much more than previous methods? We have tried to deal with this understandable sentiment by including a motivational section in which we outline the reasons why we believe that the time is ripe for a major change in the way we think about chemical bonding. However, the compelling argument in favor of adopting the "language" which we propose is actually embodied in a series of papers which follow this publication and in which we apply the newly developed theory to diverse problems in a way which makes evident that our previous "understanding" of chemistry has been often illusory.

The last statement needs some amplification. In the course of this presentation and, much more so, in following papers we will discover that previous interpretations based on qualitative MO theory of the Frontier Orbital One-Electron Perturbation MO type were either deficient or incorrect. As former practitioners of such a brand of qualitative theory, we can state that it was exactly these failures which led us to the development of the theory outlined in this article. In fact, the involvement of this author with quantum chemistry has been the direct result of the stimulating influence of the Frontier Orbital idea of Fukui and the conservation of MO symmetry idea of Woodward and Hoffmann.

Thus, we shall point out failures of qualitative MO theoretical models cognizant of the evolutionary nature of science and appreciative of the past contributions of brilliant investigators which actually brought us to the threshold on which we step now.

Finally, I wish to mention the fact that the work described here has been carried out without the support of private or Federal U.S. agencies and express my appreciation for the assistance provided during different times by Dr. James Larson and Mr. Hugh Eaton.

Nicolaos D. Epiotis

Contents

PART I

QUALITATIVE VALENCE BOND THEORY OF MODEL SYSTEMS

Introduction

"Supression of details may yield results more interesting than a full treatment. More importantly, it may suggest new concepts. Pure quantum mechanics alone, in all its details, cannot supply a definition of, e.g., an acid or a base or a double bond." These statements are attributed to E. Schrödinger and they constitute one of the earliest realizations of the necessity of interplay between "quantitative" and "qualitative" quantum theory.[1] In the former case, a preoccupation with the physical significance of mathematical expressions is secondary to obtaining a numerical answer which can be compared with the result of an experiment. Indeed, the mathematical structure of quantum mechanics itself is the actual model. By contrast, "qualitative" theory attempts, through computational tests and reference to experimental facts, to simplify the rigorous equations of "quantitative" theory so that some approximate physical model, which can be routinely applied to chemical problems without the need of explicit calculations, finally emerges. It then follows that, while "quantitative" theory can be elaborated on an *ab initio* level, "qualitative" theory is always empirical and rests on fundamental assumptions. A better "quantitative" calculation can aid the development of a better "qualitative" physical model, and vice versa. Ultimately, one hopes that the two different theoretical approaches will yield results which are in harmony between them as well as with the results of experimental studies.

We have been interested in the "quantitative" theory-"qualitative" theory-experiment triptych for about a decade during which time we have explored different theoretical frameworks and viewpoints in a variety of structure and reactivity problems.[2,3] About three years ago, our original enthusiasm and confidence in the qualitative approach began to diminish as an alarmingly large number of experimental and computational results at odds with expectations based on current qualitative theory began to appear with increasing frequency in the literature. These new facts were added to an already impressive list of "exceptions" to well known rules of qualitative theory ultimately producing a solid body of evidence which we could no longer dismiss casually or rationalize in any reasonable and self consistent manner. Immediately, disturbing questions were raised: Are the successes of qualitative theory nothing but happy coincidences? Have we developed heuristic concepts which, though frequently useful in predicting and rationalizing some (but not all) chemical trends, are based on an illusory understanding of chemical bonding? Have we been overly impressed by simplicity and have we been unwilling to tackle problems at the proper level of theory?

The above concerns have been shared by other investigators in the past and they have been expressed in the chemical literature in implicit and explicit forms. For example, a recent monograph by Schaefer[4] reviews results of _ab initio_ computations of "small" and "medium" size molecules which are not always in keeping with ordinary expectations based upon our present day qualitative understanding of chemical bonding. In addition, the mere fact that practically every theoretical interpretation of even the simplest stereochemical trend has been and still is controversial attests to a rather unclear, if not inadequate, understanding of the nature of the chemical bond. Thus, the factors responsible for the angular shapes of the simple triatomics H_2O and H_2S are still under scrutiny.[5] The origin of the rotational barrier in ethane is still being debated.[6] The intuitively unexpected preference of a large number of molecules for a "crowded" geometry, e.g., the greater stability of cis relative to trans-1,2-difluoroethylene, continues to provoke spirited discussions.[7] The list of current controversies related to interpretations of well established experimental facts continues _ad infinitum_. Indeed, one is tempted to adopt the posture that nature is too complicated and chemical and physical trends arise as a result of an indecipherable combination of multitudes of competing factors!

The purpose of this series of papers is to present a general qualitative model of chemical bonding founded principally on Valence Bond (VB) theory as an alternative to current qualitative Molecular Orbital (MO) theoretical models. In proposing a rebuilding of the conceptual superstructure of chemistry, we must provide ample evidence of the shortcomings of the MO method and spell out exactly how we plan to improve on it. This is done in the following section.

A. The Formal and Conceptual Difficulties of MO Theory

Quantum chemistry, as practiced today by most theoreticians, relies upon the Schrödinger equation, mathematical methods for its approximate solution, most notably the variational method[8] and the Rayleigh-Schrödinger and Brillouin-Wigner perturbation methods,[9] and the MO[10] and VB[11] recipes for the construction of the antisymmetrized molecular wavefunction.[12] It is not inaccurate to say that the vast majority of chemists have been nurtured with MO and VB theory, with the latter yielding decisively to the former in popularity in the last fifteen years or so. Accordingly, our first task is to survey briefly the various brands of MO theory and identify the formal and conceptual difficulties which hinder their application to chemical problems.

We can distinguish three different levels of MO theory:

a) Hückel MO (HMO) theory which encompasses a multitude of equivalent or related theoretical frameworks and their approximate versions. The characteristic features of the various HMO approaches are touched upon briefly below.

1. Pi HMO theory.[13] This is the conventional HMO theory for pi conjugated systems.

2. Extended Hückel MO (EHMO) theory.[14] This represents the generalization of pi HMO theory to pi as well as sigma orbitals and all valence electrons.[15] The well-known Mulliken-Walsh model of molecular structure[16] can be viewed as nothing but a diagrammatic representation of EHMO theory applied to molecular

structure problems. Similarly, the analysis of the stereochemistry of pericyclic reactions via one-electron MO or state correlation diagrams espoused by Woodward and Hoffmann as well as by Longuet-Higgins and Abrahamson[17] is founded on EHMO theory.

3. One-electron Perturbation MO (PMO) theory.[18] Under conditions which make the use of Perturbation Theory (PT) valid, one-electron PMO is equivalent to HMO theory. Its approximate version is the one-electron Frontier Orbital (FO) PMO model.[19] In this model, only the FO orbitals of two or more arbitrarily defined interacting fragments and the electrons which they contain are considered and PT is implemented only up to second order in energy. Currently, the FO-PMO model is the most popular qualitative theoretical tool.[20]

4. The one-electron Second Order Jahn-Teller (SOJT) model.[21] This is equivalent to a one-electron FO-PMO theory of molecular distortion. We now continue with higher level theoretical approaches.

b) Semiempirical[22a-c] and ab initio[23] monodeterminantal Hartree-Fock Self Consistent Field (SCF) MO theory, henceforth referred to as SCF-MO theory.

c) Semiempirical[22d] and ab initio[23b] polydeterminantal SCF-MO theory, henceforth referred to as SCF-MO Configuration Interaction (CI) theory. With this overview of MO theory in mind, it is not an exaggeration to say that the conceptual superstructure of organic chemistry is founded on HMO theory, with the term "HMO theory" being inclusive of all complete and approximate one-electron theories and models.

What is the reason behind the extreme and undeniable popularity of HMO theory and related qualitative theoretical models? The answer is straightforward: HMO theory is simple enough to be comprehended by the practicing chemist who does not strive to become a theoretical expert, yet hopes to become sufficiently knowledgeable in theory so that he can grasp and apply simple quantum chemical concepts and carry out explicit quantum chemical calculations with so called "canned" computer programs.[24] An undeniable impetus to this tendency has been provided by the apparent successes of qualitative HMO theory in the form of the HMO theory of pi conjugated systems,[13b,25] the Woodward-Hoffmann rules for pericyclic reactions,[26] the FO-PMO model of molecular structure and reactivity,[20] etc. Perhaps there is no better illustration of the dominant influence of HMO theory on chemistry as a whole other than the fact that, <u>in an age when sophisticated quantum chemical computations are reported in the literature with an ever increasing frequency, their interpretation is still performed by falling back on concepts founded on HMO theory, such as "aromaticity",[27] "hyperconjugation",[28] etc</u>.! Indeed, we can say that HMO theory "touches" to a smaller or greater extent every chemist, whether theoretician or experimentalist.

Simplicity is the virtue of HMO theory. What are its drawbacks? These are numerous and they can be categorized into formal and conceptual drawbacks. The formal limitations of HMO theory are well known.[29] At this level of theory, the following interactions are <u>explicitly</u> neglected:

a) "Classical" interelectronic coulomb repulsion.

b) "Classical" internuclear coulomb repulsion.

c) "Classical" coulomb attraction between an electron on one center and nuclei of different centers.

If we symbolize AO's by lower case letters, e.g., r,s,t,u, etc., and nuclei by capital letters, e.g., A,B, etc., the three approximations stated above can be articulated in mathematical language as follows:[30]

$$< rs \left| \frac{1}{r_{12}} \right| tu > = 0 \qquad (1)$$

$$\frac{Z_A Z_B}{r_{AB}} = 0 \qquad (2)$$

$$< t_A \left| \frac{Z_B}{r_{1B}} \right| t_A > = 0 \qquad (3)$$

Z_A is the effective nuclear charge of A. A special brand of HMO theory is HMO theory with neglect of AO overlap, i.e.,

$$< t | u > = 0 \qquad (4)$$

Parametrization effectively introduces some component of these effects in an implicit manner but it cannot remove the basic deficiencies of the method.

At this point, we open a parenthesis in order to specify the meaning of "classical", "semiclassical", and "nonclassical". "Classical" terms are potential and kinetic energy terms which have a counterpart in classical physics. Specifically, "classical" potential energy terms arise as a result of the interaction of electrons and nuclei in all possible ways and in a manner consistent with Coulomb's Law. Similarly, "classical" kinetic energy terms arise as a result of the motion of electrons and nuclei in a manner consistent with expectations based on classical kinematics. The "classical" terms describe interactions of "local" particle distributions, e.g., the attraction of an electron in one AO by a nucleus of some other atom. "Semiclassical" terms are analogous to "classical" terms, the only difference being that they describe interactions of "overlap" particle distributions, e.g., the attraction of an electron contained in two overlapping AO's by the nuclei of the two corresponding atoms and/or the nuclei of other atoms. Finally, "nonclassical" terms are potential and kinetic energy terms which arise as a result of the determinantal form of the total molecular wavefunction. As we shall see, a mathematical definition of these various terms can be given within the framework of VB theory. Since an MO wavefunction can be expanded to a VB wavefunction, terms defined on the basis of VB theory are equally useful in MO theory. The general problem of "translating" from MO to VB theory, and vice versa, will be taken up in the next section.

The conceptual limitations of HMO theory have been given far less attention than the formal limitations of the same theory. In fact, they are not peculiar to HMO theory but they are characteristic of MO theory, in general. Specifically, a unique analysis and interpretation of the MO wavefunction is not feasible because it is not possible to define either a unique frame of reference or an arbitrary frame of reference which, by convention, can be adhered to on a universal basis. As a result, a given MO wavefunction can be interpreted in a number of equivalent yet apparently different ways. In turn, this precludes the development of a general and self consistent theory of chemical bonding which cuts across interdisciplinary barriers. For example, consider the problem of the pi electronic structure of 1,3 butadiene. We adopt a "Molecules in Molecules"[31] approach and seek to generate the total MO wavefunction by a linear combination of fragment MO wavefunctions. Four different dissection modes which define four different types of fragment MO basis sets are shown below.

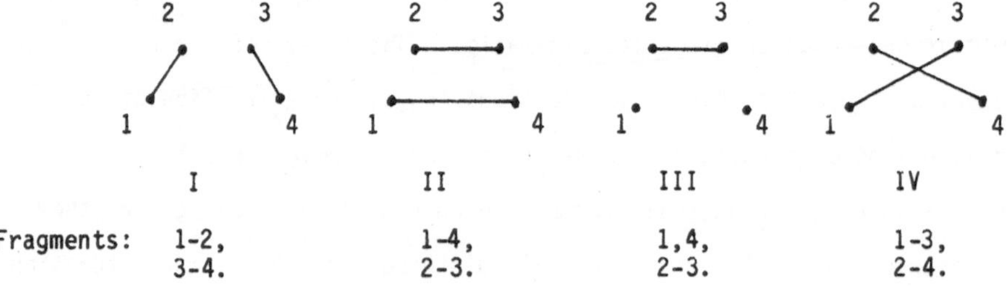

We can now generate four apparently different qualitative theories of pi electronic structure depending upon our choice of dissection mode which, nonetheless, will be equivalent at the limit of a complete variational or high order perturbational treatment. The scientific literature is replete with examples of apparently different yet equivalent treatments. In the case at hand, dissection I is a popular dissection associated with simple PMO analyses of pi conjugated systems. Dissection II is the one which led Hoffmann and coworkers to formulate the simple PMO concepts of "through bond" and "through space" interaction.[32] Dissection III has been used by Inagaki _et al_ in connection with a Linear Combination of Fragment Configurations MO-VB type theory in a recent paper entitled "Orbital Interactions in Three Systems".[33] Finally, dissection IV constitutes the dissection mode implicit in band theory.[34] In departing this subject, we emphasize once again that the conceptual limitations spoken of above are independent of the formal assumptions of the particular brand of MO theory employed, i.e., they persist at the _ab initio_ SCF-MO-CI as well as the HMO level of theory.

A limitation implies ultimately occasional or frequent failures. The following discussion is aimed at exhibiting some key failures of HMO theory of which the reader may not be aware. The specific illustrations have been chosen in such a manner so that quantitative theorists, qualitative theorists, experimentalists with a keen interest in theory, and chemists of every persuasion can find something of interest.

First, let us dispense with a well known formal deficiency of the HMO method. Specifically, the electronic structure of diradicals, i.e., systems wherein two electrons must be placed in two orbitals which are degenerate in a one electron sense, cannot be analyzed by the HMO method. In diradicals, the orbital occupancy gives rise to three singlet and one triplet low lying states which differ in the amount of interelectronic repulsion. These four states can become artificially degenerate at the level of HMO theory which neglects "classical" coulomb effects, including interelectronic repulsion. A correct description of the low lying states of diradicals can only be obtained at the level of SCF-MO-CI theory, i.e., at a level of theory which gives a proper account of electron repulsion. Qualitative discussions of the electronic structure of prototypical organic diradicals based on MO theory abound[35] and a recent review of computations of such systems has appeared.[35c]

Let us now consider additional failures of the HMO method which pertain to fundamental chemical problems and which are not widely known to practicing chemists:

a) HMO theory fails to predict correctly the energy order of the excited states of a molecule. A typical example is the energy ranking of the low lying excited states of 1,3 butadiene. If we denote the four pi MO's as π_1, π_2, π_3 and π_4, in order of increasing energy, HMO theory predicts that the lowest energy excited state will be the "singly excited" $\pi_1^2 \pi_2^1 \pi_3^1$ type state with the "diexcited" $\pi_1^2 \pi_3^2$ type state lying significantly above it. By contrast, computations at the SCF-MO-CI level show that the "diexcited" valence state lies actually below the "singly excited" state which, in fact, has Rydberg character.[36] This trend persists in longer chain linear polyenes.[37]

b) Diradicals are present either as transition states or high energy minima on the ground surface of a thermally "forbidden" reaction. In addition, they can be global minima on an upper surface whereupon part of a photochemically "allowed" reaction occurs. Solvent and substituents can change the relative energies of the four diradicals states (three singlet and one triplet) in a way which alters fundamentally the morphology of energy surfaces and, consequently, the mechanism of the thermal or photochemical reaction. Thus, there exists no unique mechanism for thermally "forbidden" or photochemically "allowed" reactions but rather a continuum of mechanistic possibilities determined by the reaction phase and the electronic structure of the reactants. A full discussion of these issues can be found in a recent monograph by one of the authors.[3d] These predictions cannot be arrived at on the basis of HMO theory due to its formal limitations.

c) HMO theory fails to treat properly multicentric chemical reactions which can be classified as thermally "allowed" and photochemically "forbidden" in the Woodward-Hoffmann sense. This point can be illustrated by the two thermal reactions shown below. In both cases, the supermolecular complexes, hexagonal H_6 and linear H_3^-, are predicted to be energy minima by HMO theory. By contrast, SCF-MO-CI computations indicate that they constitute transition states.[38,39]

$$3H_2 \longrightarrow [H_6] \longrightarrow 3H_2$$

$$H_2 + H:^- \longrightarrow [H\cdots H\cdots H]^- \longrightarrow H:^- + H_2$$

The SCF-MO-CI results are judged reliable given the fact that the benzene-like complex of 1,3 butadiene and ethylene is a transition state in the Diels-Alder reaction rather than a stable complex.[40]

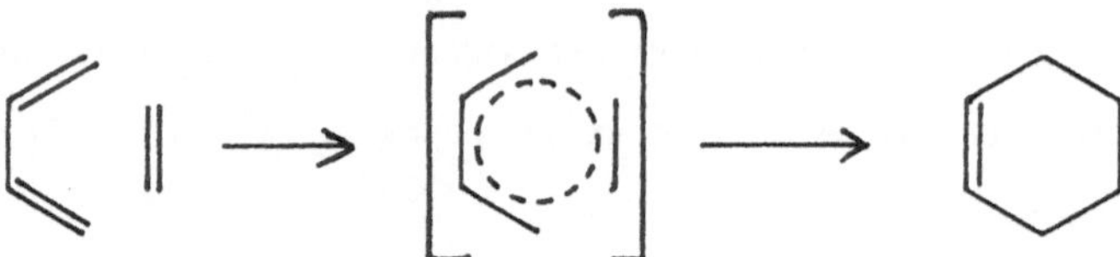

Finally, computational results as well as chemical facts[41] raise doubts as to whether a thermally "allowed" reaction occurs via a symmetrical transition state as predicted by HMO theory.

c) HMO theory fails to predict correctly the dependence of reaction stereochemistry on the electronic nature of the reactants. This is a particularly disturbing failure because one of the key aims of qualitative theory is, after all, the rational prediction of how structural modification may affect the ground or excited state geometry and reactivity of a substrate. It is also a failure which must be carefully defined if justice is to be done to HMO theory. The following discussion illustrates these problems.

It has been shown that a large number of ground state chemical trends can be rationalized by usage of the concept of aromaticity, or, equivalently, the Woodward-Hoffmann rules. Now, exceptions abound and they can be classified as "legitimate" and "illegitimate" if one focuses on the spirit rather than the letter of these theories. Specifically, the preference of a system for a least motion path over a non least motion path, when the latter is predicted to be the more favorable of the two by the Woodward-Hoffmann rules can be due to "classical" coulomb repulsion (neglected at the level of HMO theory) and/or overlap repulsion of the sigma frameworks of the two reactants (assumed to play no role in the reaction of pi systems). Such exceptions can be called "legitimate" because there exists a credible alibi, i.e., "steric effects". That is, in spirit, the Woodward-Hoffmann rules for ground state reactions are stated with the proviso that any effects not considered explicitly at the level of theory employed are constant for any two stereochemical arrangements being compared. A breakdown of this assumption can be easily forecast and it actually occurs quite frequently. In some of our past work in this area, we have identified substitution patterns which promote the occurence of "legitimate" exceptions, i.e., we have suggested strategies for narrowing the "allowed"-"forbidden" energy gap so that "steric effects" can ultimately render the "forbidden" more favorable than the "allowed" reaction.[3]

A vastly more important set of exceptions are the "illegitimate" ones, i.e., those for which a reasonable alibi cannot be found. Such exceptions signal the possibility that the Woodward-Hoffmann rules as we know them are actually a subset of a complete set of bonding selection rules which remains to be discovered. Experimental results which may illustrate "illegitimate" breakdowns, at least in the opinion of the authors, are given below.[40,42,43,44]

X=F, Δ ← (butadiene + $X_2C = CX_2$) → X=H, Δ

$F^- +$ A – N^+ (Retention) ← A=Si, Δ — A – F + $\overset{+}{N}$: — A=C, Δ → N^+– A $+F^-$ (Inversion)

In both cases, substrate modification drops the "forbidden" below the "allowed" transition state in the absence of any apparently adverse effects operating against the latter. By contrast, HMO computations predict that 4 + 2 cycloaddition and nucleophilic substitution by inversion are the preferred sterochemical reaction modes regardless of the nature of X and A.[45]

In problems of ground state molecular stereochemistry, the qualitative predictions of HMO theory stand a good chance of being "right for the right reason". This expectation is based on a large number of theoretical computations which seem to indicate that ground state stereochemistry can be predicted reasonably well by monodeterminantal MO theory.[46,47] We now pose the following question: Can we identify experimental or ab initio SCF-MO-CI computational data pertaining to problems within the domain of ground state molecular structure, i.e., the domain wherein HMO theory is expected to have a maximal chance for success, which cannot be rationalized by any concepts founded on HMO theory itself? The answer is affirmative and what follows is a brief list of examples which illustrate that our understanding of chemical bonding is primitive even at the lowest level of theoretical complexity:

a) In the last fifteen years or so and through the independent work of many investigators, a set of rules for the prediction of the stereochemical dependence of hyperconjugation has emerged.[48,49] These rules have been developed on the basis of the FO-PMO model, an approximate version of HMO theory. A typical illustrative example is given below:

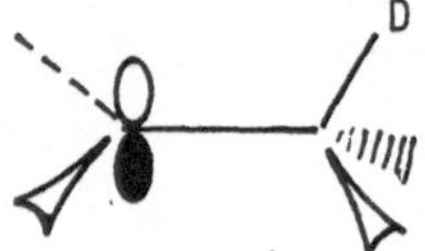

D = sigma donor

$p - \sigma_{CD}$ Stabilization
Strong Hyperconjugation

No $p - \sigma_{CD}$ Stabilization
Weak Hyperconjugation

However, while the number of successful applications is impressive, intriguing "illegitimate" exceptions to the hyperconjugation selection rules also abound. A very interesting example is provided by the comparison of conformational preferences of O_2X_2 and N_2X_4 systems, where X is an atom or group.

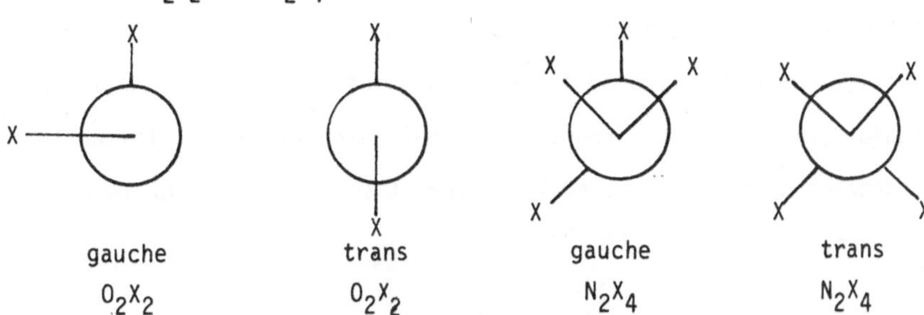

Thus, H_2O_2 is known to exist primarily in a gauche form,[50] a geometry which allows the delocalization of an oxygen lone pair to the vicinal O-H antibond. Replacement of H by F is expected to make the delocalization mechanism even more important, a prediction borne out by experiment which demonstrates that fluorination increases the preference of gauche over trans, shortens the O-O bond, and decreases the dihedral angle.[51] By contrast, N_2H_4 is also gauche,[52] presumably for the same reasons as O_2H_2, but now the perfluoro derivative exists as a mixture of gauche and trans forms[53] with the latter actually being slightly more stable than the former.[53e] In addition, the N-N bond is longer in N_2F_4 than in N_2H_4. The same trends are encountered in P_2X_4.[54,55]

b) There is a "complexity paradox" in our present ability to rationalize chemical behavior. Thus, while, for example, the stereochemistry of thermal "allowed" reactions of large molecules is understandable in terms of HMO theory based concepts, the reasons why the simple triatomic H_2O is bent and H_2S even more so continue to be matters of controversy.[5] What is the electronic origin of "symbiosis," i.e., the tendency of ligands to aggregate on one center rather than distributing themselves over more than one center?[56] Why is $CH_2 = O$ more stable than C=OH_2 while C=NH_2^+ is more stable than CH_2=N^+?[57] Why is H_2C=C=CH_2 linear but O=C=C=C=O bent?[58] No simple and satisfactory answers to these questions exist.

c) Recently, Schleyer and his coworkers have made imaginative use of ab initio SCF-MO computations in their explorations of molecular structure. Specifically, they have compared the structures of hydrocarbons and those of their lithio derivatives and demonstrated that molecules of the latter type adopt geometries which are completely different from those of the former type.[59] A typical example is given below.[60]

		X - C≡C - X	C═C bridged by X (above and below)
Relative	X = H	0	196.9
Energies: (kcal·mole^{-1})	X = Li	20.5	0

In summary, Schleyer and his group have clearly shown that there exists a conceptual vacuum in the chemistry of molecules which contain highly electropositive atoms. Why is there such a difference between C_2H_2 and C_2Li_2? How can we rationally design molecules which violate our "normal" stereochemical expectations? Unlike previous cases, where, at least, we could make a reasonable yet unsuccessful attempt towards rationalization, we are now confronted with truly novel problems which we do not even know how to approach in an a priori sense using qualitative HMO theory.

While some of the formal deficiencies of HMO theory are well publicized and their origin well understood, the conceptual problems which arise when one attempts to interpret the HMO wavefunction and, in general, any MO wavefunction are not equally appreciated. Thus, for example, there are ground state stereochemical problems where HMO theory has formally a "fighting chance" but where the interpretation of the HMO wavefunction requires skill and experience not possesed by many other than the experts of the field. An example will serve to illustrate this essential point.

Let us consider two different isomers, A and B, and ask the question: Why is A favored energetically over B? In the transformation A⟶B, some MO's are raised and some are lowered in energy. If there are n doubly occupied MO's, one may in principle group them in sets so that the energy change due to n-m MO's is nearly zero while the energy change due to m MO's parallels the total energy change. One can then claim that he has isolated a "chemical effect" in the energetic behavior of these MO's. But, if there is another way of grouping the MO's into two sets which fulfill the conditions specified above, one can equally well claim a different electronic origin of the same chemical phenomenon. Of course, there can never be universal agreement as to what should be considered constant and what should be deemed variable because the orbital manifolds change drastically from problem to problem. Hence, we are left with an abundance of potentially correct interpretations, all sounding different and none qualifying as a unique resolution of the problem. This unfortunate situation is made worse at the level of SCF-MO theory where orbital energies do not add up to the total energy. Finally, the situation becomes hopeless at the level of SCF-MO-CI theory. An excellent illustration of the conceptual difficulties of the "total molecule" HMO approach can be found in the work of Lowe who showed that more than one MO interpretation of the ethane rotational barrier is possible.[61]

The above discussion of failures and difficulties of HMO theory has been prompted by a suspicion that the reluctance of most chemists to adopt a conceptual modus vivendi which is more complex than that necessitated by HMO theory is the result of a general awareness of "how many things qualitative HMO concepts get right" and an unfortunate unawareness of "how many things qualitative HMO concepts get wrong". It is hoped that the facts presented in this section will convince the reader that there is good reason for seeking to replace qualitative HMO theory by some other superior brand of qualitative theory. Next, we consider the formal and conceptual problems associated with SCF-MO theory.

HMO theory (with overlap) contains the "semiclassical" and "nonclassical" effects which play a key role in determining whether two atoms, fragments, etc., will interact in a bonding or antibonding manner, but it neglects "classical" coulomb effects which act as modulators of the "semiclassical" and "nonclassical" effects (vide infra). On the other hand, ab initio SCF-MO theory has the formal advantage of containing in an explicit form all effects. Accordingly, one expects to generate concepts on the basis of ab initio SCF-MO theory which reach beyond those generated on the basis of HMO theory. Indeed, attempts to recast HMO based models in terms of ab initio SCF-MO theory have been made. Thus, the SCF-MO basis of the original Mulliken-Walsh model has been discussed by various authors,[62] SCF-MO perturbation schemes,[63] some inspired by the highly successful PMO model,[64] have been published, etc. Despite such efforts, SCF-MO theory never became the springboard of truly novel ideas which go beyond those inspired by HMO theory. This is due to the fact that, at the level of SCF-MO theory, a mathematical representation of the interplay between "nonclassical", "semiclassical", and "classical" effects can not be easily obtained. These effects are intermingled in a manner which does not allow an easy dissection of their mutual influence on one another. This point can be illustrated by reference to the SCF-MO total energy partition scheme first proposed by Allen[65] and, subsequently, adopted by various authors.[66] According to this treatment, the total energy, E_T, is decomposed as follows:

$$E_T = \text{Electron Kinetic Energy} + \text{Electron-Nucleus Attraction} + \text{Electron-Electron Repulsion} + \text{Nucleus-Nucleus Repulsion}.$$

This partition not only fails to project the interplay of "classical", "semiclassical", and "nonclassical" effects but it also obscures the all important "semiclassical" and "nonclassical" effects which are actually distributed among all four terms of E_T. These drawbacks are only partly present is the more informative energy decomposition scheme proposed by Morokuma and his coworkers[67] and subsequently utilized in a number of interesting studies by the orginators[68] as well as other investigators.[69]

Our inability to develop a qualitative understanding of the interplay between "classical" and "nonclassical" effects at the level of SCF-MO theory is basically the reason behind the survival and continuing popularity of HMO theory. For example, oftentime the results of SCF-MO computations are at variance with those of HMO computations. Sometimes this is due to the fact that SCF-MO theory contains effects which are absent in HMO theory. With the benefit of hindsight, we can provide an example of the latter type. Thus, both linear FHF^- and hexagonal H_6 are energy minima at the level of EHMO theory with the former remaining an energy minimum[39] and the latter becoming a saddle point at the level of SCF-MO theory.[38] Here, we know that somehow neglect of "classical" coulomb effects in the HMO method is responsible for these trends but we are unable to pinpoint how their explicit inclusion in the SCF-MO method alters the predictions of HMO theory.

While SCF-MO theory does treat explicitly "classical" coulomb interaction it does so poorly because of its monodeterminantal aspect. Thus, "covalent" bond dissociation and weak "covalent" bonding cannot be properly described at the SCF-MO level because the form of the MO wavefunction itself introduces constraints which cause ground states to be unduly "ionic" and higher excited states to be unreasonably "covalent". This deficiency is corrected at the SCF-MO-CI level.[70] However, at this level of theory, the same conceptual problems which plague us at the level of SCF-MO theory are futher aggravated.

<u>The above discussions make clear that a comprehensive qualitative theory of the chemical bond cannot be founded on MO theory.</u> Clearly, we must seek some other theoretical formulation which allows formal correctness to go hand in hand with conceptual accessibility. Valence Bond theory is an excellent choice for the following important reasons:

a) Rigorous VB theory can be formulated in a number of equivalent ways. Some of them make an unassailable solution of a problem possible at the qualitative level, i.e., the equations can be cast in such a form that a qualitative prediction can be made which is based on an explicit consideration of all "effects". This is far from possible in qualitative MO theory. Furthermore, approximate VB methods can be formulated which have a conceptual advantage over the <u>equivalent</u> approximate MO methods though they yield the same results.

b) VB theory can provide the frame of reference for the comparison of two different brands of MO theory as well as one brand of MO theory with one brand of VB theory. This can be achieved by recasting a given MO theory in VB terms and evaluating its performance by recourse to mathematically well defined VB concepts. This is an extraordinarily significant feature of VB theory since improper interpretations of chemical phenomena must be decisively rejected only after an unambiguous demonstration of their shortcomings has been presented.

c) VB theory guarantees the development of new concepts by virtue of the fact that it can portray the interplay of various "effects" in an explicit and pictorial fashion. For example, the interplay between "semiclassical" delocalization, represented by the interaction of VB "structures," and "classical" interelectronic repulsion, measured by the energy difference, at $r = \alpha$, of the VB "structures", can be easily visualized in the case of the two electron bond of H_2 from mere inspection of the "resonance hybrid":

$$\mathrm{H{:}^{-}\ H^{+}} \longleftrightarrow \mathrm{H\cdot\ \cdot H} \longleftrightarrow \mathrm{H^{+}\ H{:}^{-}}$$

We say that the high energy of the "ionic structures", due to severe interelectonic repulsion, prevents them from mixing strongly with the "covalent structure", thus restricting delocalization, and, by extension, affecting the bond energy. The fact that undergraduate and graduate education in organic chemistry is still founded on "resonance theory" is nothing but a reflection of the tremendous advantage of VB theory in giving a pictorial account of how the energies of the "structures" affect their interaction. We recall that neither SCF-MO nor SCF-MO-CI theory can be easily adapted to a qualitative theory because the interplay of "effects" is simply hidden behind the mathematical formalism of these theories.

B. Levels of VB and MO Theories

Given the present preeminence of MO theory, qualitative as well as quantitative , there is an unavoidable tendency to forget how frequently we actually revert to VB concepts when we discuss complicated systems or attempt to isolate the physical origin of a chemical phenomenon. Thus, the experimentalist still finds "arrow pushing" a useful tool in devising new synthetic routes or formulating mechanisms involving large molecules. Even more telling is the fact that in exploring the electronic nature of excited states and chemical intermediates one realizes the conceptual convenience of "radicaloid" and "zwitterionic" states.[35a] In attempting to understand how CI modifies the monodeterminantal MO wavefunction, one begins to appreciate the precarious balance of "covalent" and "ionic" terms.[71] Finally, even in ground state stereochemical problems, the appeal of the "bond" construct is hard to resist and many papers have been published which make intuitive use of VB theoretical methodology.[72] The purpose of this two-part work is to tailor VB theory into a potent intelligible, and formally correct qualitative theory of chemical bonding which can be applied uniformly to problems in all subdisciplines of chemistry. That this development is a natural and unavoidable occurence within the framework of scientific evolution can only be appreciated if one has a clear understanding of diverse theoretical approaches and their interrelationships. Hence, this section is devoted to a discussion of the various brands of MO and VB theories which exposes common denominators and fundamental differences.

The pantheons of MO and VB theories are depicted in Figure 1. Within the domain of conventional VB theory, the "perfect" eigenstates of a given system are obtained by means of a variational treatment of a trial wavefunction which is a linear combination of Configuration Wavefunctions (CW's) constructed from a nonorthogonal Atomic Orbital (AO) basis:

$$\Psi_{VB} = \lambda_1\Phi_1 + \ldots + \lambda_i\Phi_i + \mu_1\Phi_1' + \ldots + \mu_j\Phi_j' + \nu_1X_1 + \ldots + \nu_kX_k \tag{5}$$

VB - Type Theory	MO - Type Theory	Approximation
VB or HL(delocalized AO's)	SCF - MO - CI	None
HL(localized AO's)	—	Truncation
—	3 x 3 SCF - MO - CI	Truncation
—	SCF - MO	Constraint
NDO - VB	—	Integral
—	NDO - SCF - MO	Integral and Constraint
EHVB($S \neq 0$)	EHMO($S \neq 0$)	Integral and Constraint
HVB ($S = 0$)	HMO ($S = 0$)	Integral and Constraint

Figure 1. Approximate and rigorous types of Valence Bond and Molecular Orbital theories. Theories within a row are equivalent.

The CW's can be grouped into three sets:

a) One set containing all linearly independent, lowest energy "covalent" CW's which place one electron per AO to the extent possible. These CW's, Φ_i, will be referred to as Heitler-London (HL) CW's in recognition of the fact that the original Heitler-London VB theory was formulated over such a basis of CW's.

b) A second set containing all linearly independent "covalent" CW's, Φ_i', which have higher energy than the HL CW's.

c) A third set containing all linearly independent "ionic" CW's, X_i, i.e., all linearly independent CW's which couple two electrons within one and the same AO thus creating AO "vacancies" and AO "double occupancies" which coexist. Using the simple diatomic H_2 as an illustrative example, we say that the H• •H CW is the HL CW while the H^+ :H^- and H^-: H^+ CW's are the "ionic" CW's.

It must be emphasized that one may calculate electronic states by VB theory using either the variational or the perturbation methods. We shall refer to the former approach simply as the VB and the latter as the PVB approach. In developing the concepts of VB theory we shall uniformly assume that these are based either on VB theory or high order PVB theory since it is in the limit of high order perturbation that VB and PVB theories become equivalent.

Within the domain of MO theory, the "perfect" eigenstates of a given system are obtained in two steps. First, the MO eigenfunctions are obtained by means of a variational treatment of a trial wavefunction which is a Linear Combination of Atomic Orbitals (LCAO-MO's). Subsequently, the "perfect" eigenstates are obtained by means of a variational treatment of a second trial wavefunction which is now a linear combination of CW's constructed from the MO basis determined in the first step. The overall MO procedure is equivalent to the VB procedure and the effective trial wavefunction of MO theory can be written as follows:

$$\Psi_{MO} = \lambda_1\Phi_1 + \ldots + \lambda_i\Phi_i + \mu_1\Phi_1' + \ldots + \mu_j\Phi_j' + \nu_1 X_1 + \ldots + \nu_k X_k \quad (6)$$

Using the simple diatomic H_2 as an illustrative case, we note that the variational treatment of the LCAO trial wavefunction leads to the generation of two MO's, σ and σ^*. The variational treatment of a trial wavefunction which is a linear combination of the CW's σ^2, $\sigma^1\sigma^{*1}$, and σ^{*2}, where the superscript indicates the number of electrons occupying a given MO, leads to the final eigenstates. However, each MO CW is a linear combination of VB CW's in the manner indicated below. Thus, we have effectively done nothing other than perform a VB treatment via a circuitous route.

$$\sigma^2 \propto [H\cdot \ \cdot H] + \lambda\ [H\colon^- \ H^+ + H^+ \ \colon H^-] \tag{7}$$

$$\sigma^1\sigma^{*1} \propto [H\colon^- \ H^+ - H^+ \ \colon H^-] \tag{8}$$

$$\sigma^{*2} \propto [H\colon^- \ H^+ + H^+ \ \colon H^-] - \lambda\ [H\cdot \ \cdot H] \tag{9}$$

In summary, the two "perfect" theories of chemical bonding which make use of a variational treatment of the trial wavefunction are the VB and MO (commonly referred to as MO-CI) theories. At the limit of complete AO and CW basis sets, the VB and MO-CI theories yield the same answer. In this series of papers, we shall not be concerned in any way with the relative merits of the two brands of theory insofar as computational efficiency is concerned. For these matters, the reader is referred to the important work of a number of investigators who have sought to develop quantitative theory capable of handling molecular systems of interest to the practicing chemist.[73] For the purpose of future discussion, we shall refer to any variational or perturbational VB or MO-CI theory which approaches as closely as possible the "truth" as a "state of art" theory. We aspire to produce a qualitative theory which operates, albeit in a non-numerical fashion, at the same level as "state of the art" theory.

At the present time, the concepts of chemistry are founded not on "perfect" theory but, rather, on approximate theory and, more precisely, on qualitative approximate theory. There are three fundamental appoximations which are frequently used in connection with MO and VB theory as shortcuts to laborious and/or expensive computations or as simplifications necessary for the development of succinct and operationally useful qualitative arguments:

a) The truncation approximation. According to this approximation, an incomplete AO basis is used as a starting point for VB or MO computations. The problems arising from usage of an incomplete AO basis have been discussed by various authors[74] and will not concern us in this work. On the other hand, the consequences of the truncation of the CW basis set merit particular attention as they are directly relevant to our pursuits and they are briefly discussed below.

The popular truncated form of the VB wavefunction is the HL wavefunction: i.e., the linear combination of the HL CW's:

$$\Psi_{HL} = \lambda_1\Phi_1 + \lambda_2\Phi_2 + \ldots + \lambda_i\Phi_i \qquad (10)$$

The HL wavefunction is a very reasonable aproximation of the complete VB wavefunction of most homonuclear molecular systems in which the contribution of "ionic" CW's is expected to be small. On the other hand, the popular truncated form of the MO-CI wavefunction is a linear combination of three MO CW's, C_1, C_2 and C_3, which correspond to the three possible assignments of two electrons in two FO's, namely, the Highest Occupied MO (HOMO) and Lowest Unocuppied MO (LUMO) of the system of interest.

$$\Psi'_{MO} = \lambda_1 C_1 + \lambda_2 C_2 + \lambda_3 C_3 \tag{11}$$

For polyelectronic systems, this wavefunction is effectively equivalent to a linear combination of "covalent" and "ionic" VB CW's with <u>constrained</u> variational parameters λ_i, μ_j, ν_k (<u>vide infra</u>).

The important thing to note is that, in polyelectronic systems, the truncated VB wavefunction is not at all equivalent to the truncated MO-CI wavefunction. Thus, for example, the former always dissociates correctly to atoms while the latter does not. This means that if we are to start anew developing qualitative concepts, HL theory must be given highest priority insofar as approximate methods are concerned. This is true because the truncation approximation is the most reasonable of all three types of approximations, for most nonpolar systems.

b) The constraint approximation. This approximation is inherent in monodeterminantal MO theory. By contrast, no such approximation is ever used in connection with VB theory. Under the constraint approximation, the MO-CI trial wavefunction is modified as follows:

$$\Psi''_{MO} = \lambda_1(\Phi_a + \ldots + \Phi'_a + \ldots + X_a + \ldots) + \lambda_2(\Phi_b + \ldots + \Phi'_b + \ldots + X_b + \ldots) + \ldots \tag{12}$$

The VB CW's within any one parenthesis are different from those within any other parenthesis and the variational parameters λ multiply sums of "covalent", Φ and Φ', and "ionic", X , VB CW's. Note that the "covalent" and "ionic" VB CW's are no longer allowed to vary independently. This constraint is the result of the mathematical form of the MO wavefunction and its consequences are catastrophic for many types of chemical problems one is apt to encounter. The problem can be illustrated by reference to the ground state monodeterminantal MO wavefunction of the diatomic A-A. The form of the doubly occupied sigma type MO dictates that the optimal ground monodeterminantal MO wavefunction must have an equal contribution from the "covalent" and "ionic" VB CW's, i.e., there is a constraint built in the trial wavefunction which does not allow a differential contribution from the two types of VB CW's. This drawback of monodeterminantal MO theory becomes increasingly significant as the A-A bond becomes weaker. A weak electron pair bond is typically the result of a small interaction matrix element connecting the "covalent" and "ionic" VB CW's. For the case at hand, this interaction matrix element takes the following approximate form.

$$\langle A\cdot \; \cdot A|\hat{H}|A^{+} \; A\bar{:}\rangle \simeq \beta_{12} \tag{13}$$

$$\beta_{12} = K(h_{11} + h_{22})\, s_{12}/2 \tag{14}$$

Where $h_{11} = \langle x_1|\hat{H}|x_1\rangle$ (15a)

$s_{12} = \langle x_1|x_2\rangle$ (15b)

In the above expressions, β_{12} is the monoelectronic "resonance integral" $\langle x_1|\hat{H}|x_2\rangle$, K is an energy constant, h_{11} is the one electron energy of x_1, s_{12}, the AO overlap integral, and x_1 and x_2 are the two overlapping AO's of A-A. The approximation of β_{12} defined by equation (14) is known as the Wolfsberg-Helmholz approximation.[75] It follows that as the overlap of the two AO's and/or ionization energy tends to zero, the mixing of "covalent" and "ionic" CW's will tend to zero. The diminishing overlap trend is reproduced in any problem involving electron pair bond dissociation and the diminishing AO ionization energy trend is reproduced in any comparison of two systems one of which contains strongly binding atoms (i.e., atoms with large negative h_{tt}'s) and another containing weakly binding atoms (i.e., atoms with small negative h_{tt}'s). Clearly, the optimal ground state wavefunction of a weakly bound A-A diatomic must have the form:

$$\Psi^{\circ} \simeq A\cdot \ \cdot A \qquad (16)$$

By contrast, the constraint approximation dictates that monodeterminantal MO theory produces a higher energy, nonoptimal ground state wavefunction of the form shown below.

$$\Psi^{\circ} \simeq .707[A\cdot \ \cdot A] + .5[A\colon^{-} A^{+}] + .5[A^{+} \ \colon A^{-}] \qquad (17)$$

The study of chemical reactions is tantamount to the study of bond dissociation. The study of the stereochemistry of ground and excited molecules held together by weak bonds is tantamount to the study of the electronic factors which determine optimal weak bonding. Many organic and most inorganic molecules are held together by weak bonds. It is evident that the constraint approximation embodied in monodeterminantal MO theory is a very drastic approximation which must be avoided at all costs.

c) The integral approximations. These approximations can be used in connection with either VB or MO theory in order to simplify the energy and overlap matrices generated by the variational treatment of the trial wavefunction. There exists three popular integral approximations:

1. Neglect of Differential Overlap, as defined by the equality below.

$$\phi_t\phi_u = 0 \tag{18}$$

It follows that on the basis of this approximation, the following equalities also hold:

$$\langle t|u\rangle = 0 \tag{4}$$

$$\langle rs|tu\rangle = 0 \tag{19}$$

This integral approximation is denoted by adding NDO in front of the appropriate theoretical label. Typical examples of NDO-MO theories are the CNDO and INDO methods developed by Pople and his coworkers and the MNDO and MINDO methods developed by Dewar and his coworkers. NDO-VB theory is the form of VB theory implicitly used by Pauling in his brilliant investigations of molecular structure.[76]

2. Neglect of "classical" coulomb effects as defined by equalities (1)-(3). This integral approximation is denoted by appending EH in front of the appropriate theoretical label. A typical example is the EHMO method. EHVB theory has neither been formulated nor applied to polyelectronic systems.

3. Neglect of differential overlap as well as "classical" coulomb interaction, denoted by appending H in front of the appropriate theoretical label. A typical example is the well known HMO theory with neglect of overlap. Once again, HVB theory has neither been formulated nor applied to polyelectronic systems.

Let us now examine the physical consequences of the integral and constraint approximations.

Electron delocalization is the physical act of electron hopping from one atomic center to another represented by the mixing of VB CW's. Accordingly, all VB or MO wavefunctions are delocalized while, by contrast, the HL wavefunction is localized. Now, the extent of delocalization in MO and VB wavefunctions depends on the strength of the interaction of "covalent" and "ionic" VB CW's. This, in turn, depends on the relative energies of the various VB CW's as well as the energy and overlap matrix elements, H_{ij} and S_{ij}, connecting them. Different approximations have different effects on these critical quantities. Accordingly, we can view approximate quantum chemical theories as approximate descriptors of electron delocalization within a given system. More specifically, the integral approximations defined above lead to wavefunctions wherein electron delocalization is unrestrained, i.e., the mixing of the "covalent" and "ionic" VB CW's is exaggerated. The same is true for the constraint approximation.

The equivalence relationships of MO and VB rigorous and approximate theories are spelled out in Table I. Equivalency between MO and VB theory is obtained only when "classical" coulomb interaction is neglected or when CI correction is included in the MO treatment. The contents of Table I will become better understood when we discuss explicitly the physical significance of energy and overlap matrix elements over VB CW's in the following section. For the time being, we stress that this Table contains the necessary "translations" required for identifying the theoretical deficiencies of an approximate MO method via the utilization of VB concepts. For example, if one wishes to understand why HMO theory fails in a particular application, all that needs to be done is "translate" HMO to its equivalent HVB theory and identify the deficiencies of HVB theory in a pictorial

Table 1. The Correspondence of MO and VB Theories.

MO Theory	Equivalent VB Theory
HMO (S = 0)	HVB
HMO (S $\neq$ 0) or EHMO	EHVB
NDO - SCF - MO	None
NDO - SCF - MO - CI	NDO - VB
SCF - MO	None
SCF - MO - CI	Standard VB Theory. Generalized VB Theory*.

*Equivalent to $\overline{HL}$ theory defined in text.

VB sense. It is this type of approach which will allow us to place the limitations of current qualitative theoretical concepts based on HMO theory in proper perspective.

C. Qualitative Valence Bond Theory

Much of the current popularity enjoyed by MO theory, at least in organic chemistry, is due to a preparatory phase which accomplished several things. First, it familiarized many chemists, including the principal author, with basic MO theoretical methodology. Second, it demonstrated that simple concepts can be extracted from simple equations. In addition, these concepts were shown to be operationally useful, i.e., they could be used to rationalize and sometimes predict experimental results. The basic tools of MO theory have been developed by a number of pioneer authors such as Mulliken,[77] Coulson,[78] Longuet-Higgins,[79] and others.[80] Early qualitative applications of MO theory to important chemical problems were contributed by Hückel, Mulliken, Fukui,[81] and others. Finally, and quite importantly, the MO method and its applications were brought to the attention of organic chemists, by the pioneer monographs of Roberts[82] and Streitwieser. It is not inaccurate to say that the pedagogical interlude of the early 1960's paved the way to the wide acceptance of the one-electron FO-PMO concepts which now dominate the thinking of most chemists interested in the electronic structure of ground state molecules and transition state complexes.

By contrast, there has been no analogous pedagogical preparation for the acceptance of VB concepts which reaches beyond "resonance theory."[83] Thus, there is no single monograph devoted to elementary VB theory and its applications and the interested reader must find his way to the original papers of Slater, Van Vleck, Eyring,[84] Pauling,[85] McWeeny,[86] and others in order to understand the "how to do it" aspect of the theory and gain familiarity with its conceptual as well as formal advantages and disadvantages. Now, these papers have been written with different aims in mind so that the chemist who is interested in applied theory may find little encouragement in them to go on with the task of mastering VB theory. The situation is hardly improved in texts of quantum chemistry. Only some of the early texts contain complete discussions of elementary VB theory.[87] For a first exposure to VB theory, the monograph of Sandorfy[87e] is particularly clear in its exposition of bond eigenfunctions and the computation of the elements of the energy and overlap matrices over an HL CW basis. By contrast, texts of recent vintage tend to emphasize heavily MO theory and usually discussion of VB theory is left out. In this paper, we make an effort to alleviate the "VB background problem", which may unduly restrict the potential audience of this presentation, by starting from the beginning, so to speak, to the extent permitted by editorial policy. This is absolutely necessary if the reader who is not very familiar with VB formalism is to understand the subsequent development of equations which, in turn, will form the basis for a physical interpretation of VB wavefunctions.[88]

The qualitative VB treatment of any given atom, molecule, complex, etc., involves the following steps:

a) The nonrelativistic total Hamiltonian of the system is written down.

b) All possible CW's are generated by permuting all electrons among all nonorthogonal AO's of the atoms which constitute the entire system. Each CW is written as a Slater determinant or a linear combination of Slater determinants, as appropriate.

c) The trial VB wavefunction is constructed as a linear combination of CW's.

d) It is assumed that the electronic and nuclear wavefunctions are separable (Born-Oppenheimer approximation). Accordingly, one may proceed in two different ways to determine the total energy of a given system:

1. By defining an <u>electronic</u> Hamiltonian, determining the total electronic energy by use of the variation method, and adding the nuclear repulsion term in order to obtain the total energy.

2. By defining a total electrostatic Hamiltonian and determining the total energy by use of the variation method.

The first approach is the one which is most compatible with the thinking of people used to MO theory and MO computations. The second approach has distinct conceptual advantages when used in conjunction with VB theory, for reasons that will be discussed elsewhere, and it will be the method used consistently in this work, unless otherwise stated.

The total electrostatic Hamiltonian has the form shown below.

$$H = \sum_{A} \sum_{B>A} \frac{Z_A Z_B}{r_{AB}} + \sum_{i} \frac{-\nabla_i^2}{2} + \sum_{i} \sum_{A} \frac{-Z_A}{r_{iA}} + \sum_{i} \sum_{j>i} \frac{1}{r_{ij}} \qquad (20)$$

NUCLEAR REPULSION — ELECTRON KINETIC ENERGY — ELECTRON-NUCLEUS ATTRACTION — ELECTRON REPULSION

With this Hamiltonian, a variational treatment of the trial VB wavefunction leads to the standard eigenvalue problem represented by equation (21).

$$\left| \underline{H_{ij}} - E\,\underline{S_{ij}} \right| = 0 \qquad (21)$$

It should be noted that relativistic effects are neglected in this formulation and they must be incorporated, either empirically or explicitly, whenever appropriate.

e) The energy matrix $\underline{H_{ij}}$ is constructed and attention is paid to the attractive or repulsive character of each CW with respect to some assumed reaction coordinate of interest. In addition, the mode of interaction, i.e., strong versus weak, of critical CW's is identified. The overlap matrix $\underline{S_{ij}}$ is treated in the same fashion. The H_{ij} and S_{ij} matrix elements are now matrix elements between determinantal wavefunctions and much of the abhorrence of many chemists for VB theory is due to the fact that the expressions are lengthy and seemingly difficult to interpret. We shall see that, while the former criticism is valid, the latter is far from being so.

f) The energies of the final eigenstates are deduced qualitatively from knowledge of the H_{ij}'s and S_{ij}'s. We now seek to understand the factors which determine the magnitudes of the elements of the energy and overlap matrices which, in turn, determine the final energies of the VB eigenstates.

In this work, a VB CW, Φ_i is defined as the linear combination of Slater determinants, X_a, which produces a linearly independent function consistent with the occupancy of the spin orbitals and total spin as represented pictorially in the conventional way.

$$\Phi_i = \lambda_a X_a + \lambda_b X_b + \ldots \tag{22}$$

For example, the drawing shown below is associated with the unnormalized function written underneath it.

$\phi \uparrow \quad \downarrow \psi$

$$\Phi = |\phi\bar{\psi}| + |\psi\bar{\phi}|$$

The interaction matrix element, H_{ab}, over Slater determinants has the general form shown below, where $\hat{P}$ represents the electron permutation operator and p the parity (odd or even) of the permutation. The symbol d,X_a denotes the diagonal element of determinant X_a.

$$H_{ab} = \langle X_a|\hat{H}|X_b\rangle = \sum_{\hat{P}} (-1)^p \langle d,X_a|\hat{H}\mid \hat{P}\, d,X_b\rangle \tag{23}$$

The interaction matrix element, H_{ab}, can be thought of as a sum of energy terms arising from electron permutations which generate neither orbital nor spin orthogonalities. These electron permutations can be categorized as follows:

a) The "zero" permutation, P^o, which does not interchange the coordinates of any electron pair. This electron permutation is guaranteed to produce the largest single energy term if the ordering of the spin orbitals and electrons in X_a and X_b is such as to produce maximum coincidence between X_a and X_b. Henceforth, we shall define that the energy term generated by P^o is the one consistent with maximum coincidence of X_a and X_b. Furthermore, the energy term associated with the P^o permutation will be denoted by E (P^o).

b) The single permutation, P_{xy}, which interchanges the coordinates of one electron pair. With the conventions defined in (a), each of the energy terms generated by a single interchange of electron pair coordinates will be smaller than that generated by a zero interchange. However, the sum of all the energy terms generated by all single interchanges may, in some cases, exceed, in absolute magnitude, the single term generated by the zero interchange. The energy associated with a P_{xy} permutation will be denoted by E (P_{xy}) and the sum of energy terms resulting from single permutations by E (P^1), where the superscript denotes the type of electron interchange.

c) The double permutation, $P_{xy,x'y'}$, which interchanges the coordinates of two electron pairs.

The list continues ad infinitum, but the trends discussed above remain unaltered, and the notation introduced in (a) and (b) can be easily adapted to higher order permutations. We can now express the diagonal and off-diagonal matrix elements over Slater determinants in the following way:

$$H_{aa}=E_{aa}(P^{o})+E_{aa}(P_{xy})\ldots\ldots+E_{aa}(P_{xy,x'y'})\ldots\ldots+E_{aa}(P_{xy,x'y',x''y''}) \qquad (24)$$

or, by dropping the subscripts aa in the right hand of the equation,

$$H_{aa} = E(P^{o}) + E(P^{1}) + E(P^{2}) +\ldots\ldots \qquad (25)$$

$E(P^{q})$ is the sum of qth permutation terms.

$$H^{o}_{aa} = E(P^{o}) \qquad (26)$$

$$EX_{aa} = E(P^{1}) + E(P^{2}) +\ldots. \qquad (27)$$

Hence,

$$H_{aa} = H_{aa}^{o} + EX_{aa} \qquad (28)$$

The term H^{o}_{aa} represents the "classical" energy of the system and the term EX_{aa} the "nonclassical", or, the exchange energy of the X_a CW. We can further define the following quantities:

$$H_{aa}^{o} = F_{aa} + G_{aa} \qquad (29)$$

$$L_{aa} = G_{aa} + EX_{aa} \qquad (30)$$

Hence,

$$H_{aa} = F_{aa} + G_{aa} + EX_{aa} \qquad (31)$$

In a VB theoretical sense, F_{aa} represents the energy of the isolated fragments, i.e., atoms, G_{aa} represents the "classical" coulomb interaction, EX_{aa} the exchange interaction of the fragments, and L_{aa} the total interaction energy.

Similarly, we can express off-diagonal elements over Slater determinants in the following way:

$$H_{ab} = E_{ab}(P^{o}) + E_{ab}(P_{xy}).......+ E_{ab}(P_{xy,x'y'})... \quad (32)$$

or, by dropping the subscripts ab in the right hand of the equation,

$$H_{ab} = E(P^{o}) + E(P^{1}) + E(P^{2}) + ... \quad (33)$$

$$H^{o}_{ab} = E(P^{o}) \quad (34)$$

$$EX_{ab} = E(P^{1}) + E(P^{2}) +... \quad (35)$$

Hence,

$$H_{ab} = H^{o}_{ab} + EX_{ab} \quad (36)$$

In a VB theoretical sense, H^{o}_{ab} represents the "semiclassical" interaction term and EX_{ab} the exchange interaction term.

Finally, the analogous expressions for the diagonal and off-diagonal elements of the overlap matrix are the following:

$$S_{aa} = S(P^{o}) + S(P^{1}) + ... \quad (37)$$

$$S^{o}_{aa} = S(P^{o}) = 1 \quad (38)$$

$$SX_{aa} = S(P^{1}) + S(P^{2}) + ... \quad (39)$$

Hence,

$$S_{aa} = 1 + SX_{aa} \quad (40)$$

Similarly,

$$S_{ab} = S(P^{o}) + S(P^{1}) + ... \quad (41)$$

$$S^{o}_{ab} = S(P^{o}) \quad (42)$$

$$SX_{ab} = S(P^{1}) + S(P^{2}) + ... \quad (43)$$

Hence,

$$S_{ab} = S^{o}_{ab} + SX_{ab} \quad (44)$$

Our next goal is to cast each energy term, $E_{ab}(P_{xy},..)$, in a form which not only makes physical sense but which is also helpful in projecting how popular theoretical approximations may lead to erroneous conclusions. To this extent, we note that each of the exchange terms is a sum of monoelectronic and bielectronic terms which can be grouped into two sets so as to facilitate applications of VB theory to chemical problems heretofore treated by the simple HMO method with inclusion of overlap. An extraordinarily useful partition of the monoelectronic and bielectronic terms is the following:

1. All monoelectronic terms which are contained in the corresponding EHMO treatment (see section A) are grouped in one set and the corresponding sum is symbolized by T_n.

2. All terms which attain a value of zero in the corresponding EHMO treatment (see section A) are grouped into a different set and the corresponding sum is symbolized by R_n.

With the above definitions, we obtain the following relationships:

$$E(P_{xy},..) = E'(P_{xy},..) + E''(P_{xy},..) \qquad (45)$$

with

$$E'(P_{xy},..) = T_n \qquad (46)$$

$$E''(P_{xy},..) = R_n \qquad (47)$$

Also

$$EX = EX' + EX'' \qquad (48)$$

with

$$EX' = \sum_n T_n \qquad (49)$$

$$EX'' = \sum_n R_n \qquad (50)$$

Thus we can write:

$$H_{aa} = F_{aa} + G_{aa} + \sum_n (T_n + R_n) \qquad (51)$$

$$H_{ab} = H^0_{ab} + \sum_n (T_n + R_n) \qquad (52)$$

The following trends are noteworthy:

1. T_n and R_n have opposite signs.

2. T_n is much larger in absolute magnitude than R_n. This can be easily demonstrated by actual substitution of numerical quantities. Hence, we can regard R_n as the <u>attenuator</u> of T_n. For this reason, EX'' is the attenuator of EX'.

3. Each R_n is made up of monoelectronic, bielectronic, and nuclear exchange integrals. The monoelectronic and nuclear exchange terms have opposite signs and tend to cancel out. Hence, we can attribute the attenuating effect of R_n and, hence, EX", to net bielectronic exchange interaction which operates in opposition to the monoelectronic interaction in T_n. Henceforth, we shall refer to T_n and R_n as "monoelectronic" and "bielectronic" terms, respectively, in order to convey the fact that T_n behaves as a monoelectronic term of one sign and R_n as a bielectronic term of opposite sign.

In summary, the new vocabulary to be used consistently throughout this series of papers is the following:

a) $E(P_{xy},\ldots)$: Energy Exchange Term

b) $E'(P_{xy},\ldots)$ or T_n: "Monoelectronic" Energy Exchange Term

c) $E''(P_{xy},\ldots)$ or R_n: "Bielectronic" Energy Exchange Term

d) EX: Exchange Energy

e) EX': "Monoelectronic" Exchange Energy

f) EX'': "Bielectronic" Exchange Energy

At the level of HMO or EHMO theory:

$$E''(P_{xy},\ldots) = R_n = EX'' = 0 \tag{53}$$

We can treat each "semiclassical" term in a similar fashion by partitioning it into a "monoelectronic" and a "bielectronic" part, as follows:

$$H^{o}_{ab} = H^{o}_{ab}{}' + H^{o}_{ab}{}'' \tag{54}$$

In HMO or EHMO theory, we have:

$$H^{o}_{ab}{}'' = 0 \tag{55}$$

Once again, $H^{o}_{ab}{}'$ and $H^{o}_{ab}{}''$ have opposite signs, $H^{o}_{ab}{}'$ is larger than $H^{o}_{ab}{}''$ in absolute magnitude and, thus, we can regard the latter as the attenuator of the former.

Next, we turn our attention to matrix elements over CW's, recalling that each linearly independent CW, Φ_i, is written as a linear combination of Slater determinants, X_a, consistent with the orbital occupancy and total spin of the electronic configuration represented by Φ_i:

$$\Phi_i = \sum_a^k \lambda_a X_a \tag{56}$$

The diagonal matrix element H_{ii} takes the form:

$$H_{ii} = \sum_a^k \lambda_a^2 H_{aa} + \sum_a^k \sum_b^k \lambda_a \lambda_b H_{ab} \tag{57}$$

By substituting for H_{aa} and H_{ab} from equations (31) and (36), respectively, we obtain

$$H_{ii} = \sum_a^k \lambda_a^2 (F_{aa} + G_{aa}) + \sum_a^k \lambda_a^2 EX_{aa} + \sum_a^k \sum_b^k \lambda_a \lambda_b H^{o}_{ab} + \sum_a^k \sum_b^k \lambda_a \lambda_b EX_{ab} \tag{58}$$

Now, we can group the exchange and "semiclassical" terms into a single term denoted by X_i so that equation (58) becomes:

$$H_{ii} = \sum_a \lambda_a^2 (F_{aa} + G_{aa}) + X_i \tag{59}$$

or

$$H_{ii} = k'F_{aa} + k'G_{aa} + X_i \tag{60}$$

where

$$k' = \sum_a \lambda_a^2 \tag{61}$$

By incorporating k' into F_{aa} and G_{aa}, we can write the functional form of H_{ii} as follows:

$$H_{ii} \propto F_i + G_i + X_i \tag{62}$$

Next, we distinguish two types of off-diagonal matrix elements, H_{ij}. The first is defined over the Φ_i and Φ_j given below, i.e., two CW's which have common Slater determinants.

$$\Phi_i = \sum_a^k \lambda_a X_a \tag{63}$$

$$\Phi_j = \sum_a^z \lambda_a X_a \tag{64}$$

We write:

$$H_{ij} = \sum_a^k \lambda_a^2 (F_{aa} + G_{aa}) + \sum_a^k \lambda_a EX_{aa} + \sum_a^k \sum_b^k \lambda_a \lambda_b H^\circ_{ab} +$$

$$\sum_a^k \sum_b^k \lambda_a \lambda_b EX_{ab} + \sum_a^k \sum_{b=\ell}^z \lambda_a \lambda_b H^\circ_{ab} + \sum_a^k \sum_{b=\ell}^z \lambda_a \lambda_b EX_{ab} \tag{65}$$

Once again, we can collapse the exchange and "semiclassical" terms into a single term denoted by X_{ij} so that equation (65) becomes

$$H_{ij} = \sum_{a}^{k} \lambda_a^2 (F_{aa} + G_{aa}) + X_{ij} \tag{66}$$

The functional form of H_{ii} becomes:

$$H_{ij} \; \alpha F_c + G_c + X_{ij} \tag{67}$$

The script c implies F and G terms of the common Slater determinants.

The second type of off-diagonal matrix element is defined over the Φ_i and Φ_j given below, i.e., two CW's which do not have common Slater determinants.

$$\Phi_i = \sum_{a}^{k} \lambda_a X_a \tag{68}$$

$$\Phi_j = \sum_{a=\ell}^{z} \lambda_a X_a \tag{69}$$

We have:

$$H_{ij} = \sum_{a}^{k} \sum_{b=\ell}^{z} \lambda_a \lambda_b H^{\circ}_{ab} + \sum_{a}^{k} \sum_{b=\ell}^{z} \lambda_a \lambda_a EX_{ab} \tag{70}$$

or, by collapsing the exchange and "semiclassical" terms into a single term X_{ij}:

$$H_{ij} = X_{ij} \tag{71}$$

Henceforth, we shall denote by X the sum of "semiclassical" and exchange terms, in general. Accordingly, the functional dependence of each matrix element H_{ij} can now be written as follows:

$$H \alpha F + G + X \quad (72)$$

This is the most general form of the VB matrix element and it reads as follows: The energy of a given CW, Φ_i expressed by H_{ii}, as well as the interaction of two CW's, Φ_i and Φ_j, expressed by H_{ij}, depends on the sum of the following:

a) The energy of the isolated atoms or fragments represented by F.

b) The "classical" coulomb interaction energy represented by G.

c) The "semiclassical" plus exchange interaction energy X, henceforth denoted as the <u>overlap</u> <u>interaction</u>.

We note that F (as well as F_i) and G (as well as G_i), contain adjustment factors, some of which may remain or disappear depending on the type of VB theory employed and all of which must be computed appropriately in any quantitative application. The "pure" energy of the isolated atoms is F_{aa} and the "pure" "classical" coulomb interaction energy is G_{aa} [equation (29)]. Henceforth, in order to avoid excessive notation, we shall represent F and F_{aa} simply by F and G and G_{aa} by G assuming that the precise meaning of these symbols is self-evident from the context of the statement.

Finally, we can use the definition of the system of equations (45) to (50) in order to break up X into "monoelectronic" part, X', comprised of terms "contained" in EHMO theory, and a "bielectronic" part, X", made up of terms neglected at the level of EHMO or HMO theory.

$$X = X' + X''$$

This will be an important tool in the analysis of the deficiencies of current concepts.

D. Physical Interpretation of Diagonal Energy Matrix Elements

Table 2 shows six prototypical CW's which illustrate how bonding and antibonding interactions between atoms enter through one class of matrix elements, i.e., the H_{ii}'s. The definitions of the various integrals of Table 2 are given below. D and A represent the two fragments, i.e., atoms, d and a represent the corresponding effective nuclei, 1 and 2 stand for the two AO's, x_1 and x_2, and r and R are nucleus-electron and nucleus-nucleus distances, respectively.

$$s = \langle 1|2\rangle \quad \text{(AO overlap integral)} \tag{73}$$

$$\left.\begin{aligned} \varepsilon_1 &= \langle 1|-1/2\nabla^2 - Z_d/r|1\rangle \\ \varepsilon_2 &= \langle 2|-1/2\nabla^2 - Z_a/r|2\rangle \end{aligned}\right\} \quad \text{(one electron AO energy)} \tag{74}$$

$$\left.\begin{aligned} V_d &= \langle 2|-Z_d/r|2\rangle \\ V_a &= \langle 1|-Z_a/r|1\rangle \end{aligned}\right\} \quad \text{(nucleus-electron coulomb interaction)} \tag{75}$$

$$V_{da} = Z_a Z_d/R \quad \text{(nucleus-nucleus coulomb interaction)} \tag{76}$$

$$\left.\begin{aligned} J_{11} &= \langle 11|11\rangle \\ J_{12} &= \langle 11|22\rangle \end{aligned}\right\} \quad \text{(electron-electron coulomb interaction)} \tag{77}$$

$$K_{12} = \langle 12|12\rangle \quad \text{(electron-electron coulomb exchange interaction)} \tag{78}$$

$$\beta = \langle 1|-1/2\nabla^2 - Z_d/r - Z_a/r|2\rangle \quad \text{(AO "resonance integral")} \tag{79}$$

The following points merit attention:

a) As we have seen, the total energy of a VB CW can be written as follows:

$$H_{ii} = F_i + G_i + X_i \tag{62}$$

When the system under investigation is comprised of neutral atoms, the "classical" interaction term, G_i, tends to zero and remains much smaller than the X_i term, at least in most cases of interest. This can be illustrated by

Table 2. Energies of Prototypical Configuration Wavefunctions (Non-Orthogonal AO Basis)

Φ_i	Pictorial Representation	$E_i =$	Classical Terms		Overlap Terms	
			F_i	G_i	X'_i	X''_i
Φ_1	x_1 — ; ↑ x_2 ; D A	$E_1 =$	ε_2	$+V_d$ $+V_{da}$		
Φ_2	x_1 ↑ ; ↓ x_2 ; D A	$E_2 = \frac{1}{1+s^2}[\varepsilon_1+\varepsilon_2$		$+J_{12}+V_d+V_a+V_{da}$	$+2\beta s$	$+K_{12}+V_{da}s^2]$
Φ_3	x_1 — ; ↑↓ x_2 ; D A	$E_3 =$	$2\varepsilon_2+J_{22}$	$+2V_d$ $+V_{da}$		
Φ_4	x_1 ↑ ; ↑ x_2 ; D A	$E_4 = \frac{1}{1-s^2}[\varepsilon_1+\varepsilon_2$		$+J_{12}+V_d+V_a+V_{da}$	$-2\beta s$	$-K_{12}-V_{da}s^2]$
Φ_5	x_1 ↑ ; ↑↓ x_2 ; D A	$E_5 = \frac{1}{1-s^2}\Big[$	$\varepsilon_1+2\varepsilon_2+J_{22}$	$+2J_{12}+2V_d+V_a+V_{da}$	$-2\beta s_2$ $-\varepsilon_2 s^2$	$-K_{12}$ $-2\langle 22\|12\rangle s$ $-V_d s^2-V_{da}s^2\Big]$
Φ_6	x_1 ↑↓ ; ↑↓ x_2 ; D A	$E_6 = \frac{1}{[1-s^2]^2}\Big[$	$2\varepsilon_1+2\varepsilon_2$ $+J_{11}+J_{22}$	$4J_{12}+2V_d$ $+2V_a+V_{da}$	$-4\beta s$ $-2(\varepsilon_1+\varepsilon_2)s^2$ $+\ 4\beta s^3$	$-4\langle 12\|11\rangle s$ $-4\langle 12\|22\rangle s$ $-2\ J_{12}s^2-2K_{12}$ $-2(V_d+V_a)s^2$ $+6K_{12}s^2$ $(-2s^2+s^4)V_{da}\Big]$

reference to the fictitious diatomic H_D-H_A, where each pseudohydrogen has a modified nuclear charge (Z_d = 0.75 and Z_a = 1.25). By using the integrals of Table 3, we calculate the G_2 and X_2 terms of E_2 in Table 2 to be -.04 au and -.40 au. Thus, we can say that the contribution to the bonding or anti-bonding interaction of two atoms made by the diagonal matrix element, H_{ii}, is a strong function of the X_i term of H_{ii}. However, X_i is a sum of a large term, X_i', and a smaller term, X_i''. Accordingly, we can develop a physical interpretation of CW's by reference to the largest term of the corresponding H_{ii} matrix element, i.e., X_i', the "monoelectronic" overlap term.

b) The sign and magnitude of X_i' determines the nature of electron interaction within each CW. By reference to Table 2, and by recalling that 2βs is a negative energy quantity, we can identify the following trends:

1. The interaction of two singlet coupled electrons (as in Φ_2) is attractive.

2. The interaction of two triplet coupled electrons (as in Φ_4) as well as the interaction of three electrons in two AO's (as in Φ_5) is repulsive.

3. The interaction of two doubly occupied AO's (as in Φ_6) is repulsive. The repulsion is roughly twice as large as that between two triplet coupled electrons.

c) The approximate forms of VB theory and their equivalent MO versions can be nicely illustrated by reference to the VB CW Φ_2 and the associated energy expression E_2 given in Table 2. Thus, in HVB and HMO theory, the F, G, X', and X'', terms are reduced as follows:

$$F = \varepsilon_1 + \varepsilon_2 \tag{80}$$

$$G = 0 \tag{81}$$

$$X' = 0 \tag{82}$$

$$X'' = 0 \tag{83}$$

Table 3. H_A - H_D integrals,[a,b,c]

S	0.50
ε_2	-0.77
ε_1	-0.16
J_{12}	0.48
J_{11}	0.77
J_{22}	0.77
K_{12}	0.16
$<11/21>$	0.31
$<22/21>$	0.31
V_a	-0.64
V_d	-0.38
β	-0.68
V_{da}	0.50

a. Energy in a.u.'s.

b. $Z_a = 1.25$; $Z_d = 0.75$; $r_{ad} = 1.00$ Å.

c. Computed with the STO - 3G basis.

In EHVB and EHMO theory, the same three terms are reduced as follows:

$$F = (1+s^2)^{-1} (\varepsilon_1+\varepsilon_2) \tag{84}$$

$$G = 0 \tag{85}$$

$$X' = (1+s^2)^{-1} 2\beta s \tag{86}$$

$$X'' = 0 \tag{87}$$

Finally, in NDOVB theory, the same three terms are reduced as follows:

$$F = \varepsilon_1 + \varepsilon_2 \tag{88}$$

$$G = J_{12} + V_a + V_d + V_{da} \tag{89}$$

$$X' = 0 \tag{90}$$

$$X'' = 0 \tag{91}$$

The energy expressions for Φ_2 at the various levels of VB and MO theory become as shown in Table 4.

The concept of excitation energy or promotional energy is a very useful concept of quantum theory. Two different types of excitations can be conceived: Intrafragmental or Local Excitation (LE) and interfragmental or Charge Transfer (CT) excitation. In qualitative VB theoretical applications to model systems, LE is seldom an important consideration. Hence, in the space below we give a brief discussion of CT excitation as represented by the energy difference of Φ_2 and Φ_3 in Table 4.

As we have already discussed, electron delocalization is the physical act of electron hopping from one AO to another reproduced at the VB level by the mixing of two CW's which differ in AO occupancy. For example, the interaction of Φ_2 and Φ_3 of Table 2 brings about an electron jump from the x_1 to the x_2 AO of the system, i.e., it brings about delocalization from x_1 to x_2. The extent of

Table 4. Energy Expressions for Φ_2 and Φ_3 at Various Levels of Theory.[a]

	VB	HVB=HMO	EHVB=EHMO	NDO-VB
E_2 [b]	$(\varepsilon_1+\varepsilon_2)+G+2\beta s+K_{12}$	$\varepsilon_1+\varepsilon_2$	$(\varepsilon_1+\varepsilon_2)+2\beta s$	$(\varepsilon_1+\varepsilon_2)+G$
E_3	$(2\varepsilon_2+J_{22})+(2V_d+V_{da})$	$2\varepsilon_2$	$2\varepsilon_2$	$(2\varepsilon_2+J_{22})+(2V_d+V_{da})$
P_2	0	0	0	0
$P_3 = \Delta F$	$(\varepsilon_2-\varepsilon_1)+J_{22}$	$(\varepsilon_2-\varepsilon_1)$	$(\varepsilon_2-\varepsilon_1)$	$(\varepsilon_2-\varepsilon_1)+J_{22}$
$E_3-E_2 = \Delta E$	$(\varepsilon_2-\varepsilon_1)+(V_d-V_a)+\begin{pmatrix}-J_{12}+J_{22}\\-2\beta s-K_{12}\end{pmatrix}$	$(\varepsilon_2-\varepsilon_1)$	$(\varepsilon_2-\varepsilon_1)-2\beta s$	$(\varepsilon_2-\varepsilon_1)+\begin{pmatrix}V_d-V_a\\+J_{22}-J_{12}\end{pmatrix}$
H_{23}	$\sqrt{2}\begin{pmatrix}\beta+(\varepsilon_2+V_a)s+\langle 22\|12\rangle\\+V_{da}s\end{pmatrix}$	$\sqrt{2}\,\beta$	$\sqrt{2}\,(\beta+\varepsilon_2 s)$	$\sqrt{2}\,\beta$
S_{23}	$\sqrt{2}s$	0	$\sqrt{2}s$	0

[a] Expressions obtained by setting $s^2 = 0$

[b] $G = J_{12} + V_d + V_a + V_{da}$

delocalization depends on the energy separation of Φ_2 and Φ_3. At the level of VB theory, we may regard the lowest energy CW as the ground CW and all others as excited CW's. With these definitions, the energy separation of the ground CW and a given excited CW at infinite interatomic distance constitutes a very useful index of the intrinsic tendency of a system towards delocalization aside from any atomic interaction considerations. With this in mind, let us consider the ground and excited CW's, $\Phi = \Phi_2$ and $\Phi^* = \Phi_3$, respectively. We have:

$$\langle\Phi|\hat{H}|\Phi\rangle = E \propto F + G + X \tag{92}$$

$$\langle\Phi^*|\hat{H}|\Phi^*\rangle = E^* \propto F^* + G^* + X^* \tag{93}$$

$$E^* - E = (F^* - F) + (G^* - G) + (X^* - X) \tag{94}$$

$$\Delta E = \Delta F + \Delta G + \Delta X \tag{95}$$

At infinite atomic distance

$$\Delta G = \Delta X = 0 \tag{96}$$

Thus, we have

$$\Delta E = \Delta F \tag{97}$$

$$\Delta F = P^* - P \tag{98}$$

Where $P = 0$

The energy quantities P and P* are the respective promotional or excitation energies of Φ and Φ^*. We can say that Φ is a ground CW ($P = 0$) while Φ^* is a singly excited CW ($P^* = \Delta F$). In most cases of interest, the HL CW (CW's) is (are) the ground CW (CW's) and all other "ionic" CW's are excited CW's, with the degree of excitation being determined by the number of electron shifts required to transform the ground CW to a higher energy "ionic" CW. Typical examples are given below.

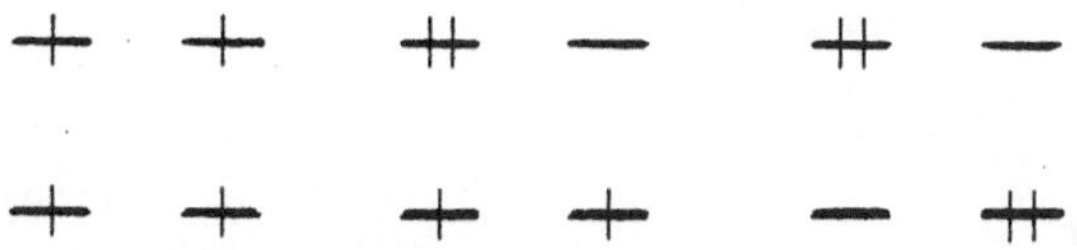

Ground	Singly "ionic" or Singly excited	Doubly "ionic" or Doubly excited

Now, in qualitative theory we are always interested in relative comparisons. Accordingly, it is useful to replace equation (62) by the equation shown below.

$$H_{ii} = P_i + G_i + X_i \tag{99}$$

This equation will often form the basis for the discussion of the H_{ii} matrix elements associated with a given type of VB theory.

E. Physical Interpretation of the Interactions of CW's:"Ionic" Versus "Covalent" Delocalization

In perturbation theory the interaction of two nondegenerate CW's, Φ_i and Φ_j, is expressed by the following equations:[89]

$$\delta E_i = (H_{ij} - E_i S_{ij})^2 / E_i - E_j \tag{100}$$

$$\delta E_j = (H_{ij} - E_j S_{ij})^2 / E_j - E_i \tag{101}$$

Clearly, the interaction depends on three key factors:

1. The energy gap separating Φ_i and Φ_j, $|E_i - E_j|$, denoted by ΔE.
2. The interaction matrix element H_{ij}.
3. The overlap integral S_{ij}.

As ΔE and S_{ij} decrease and H_{ij} increases in absolute magnitude the interaction becomes stronger.

The interaction of two degenerate CW's expressed by the equations shown below can be analyzed in a similar fashion.

$$\delta E_i = (H_{ij} - E_i S_{ij})/(1 + S_{ij}) \tag{102}$$

$$\delta E_j = -(H_{ij} - E_j S_{ij})/(1 - S_{ij}) \tag{103}$$

In the above expressions, H_{ij}, S_{ij}, and E_i are:

$$H_{ij} = \langle \Phi_i | \hat{H} | \Phi_j \rangle \tag{104a}$$

$$S_{ij} = \langle \Phi_i | \Phi_j \rangle \tag{104b}$$

$$E_i = \langle \Phi_i | \hat{H} | \Phi_i \rangle \tag{104c}$$

In a most general sense, we can say that in VB theory the interaction matrix element, H_{ij}, defines electron delocalization or electron localization of the following four major types:

a) Electron Localization. In this case, H_{ij} connects two independent CW's which describe the same orbital occupancy. We say that the mixing of CW's of this type defines electron localization. An example is given below.

$$\left[\begin{matrix} x_2 \uparrow & \downarrow x_3 \\ x_1 \uparrow & \downarrow x_4 \end{matrix}\right]_{\Phi_1} \overset{H}{\longleftrightarrow} \left[\begin{matrix} x_2 \uparrow & \uparrow x_3 \\ x_1 \downarrow & \downarrow x_4 \end{matrix}\right]_{\Phi_2}$$

b) "Covalent" Delocalization. In this case, H_{ij} connects two CW's which differ by one occupied spin orbital and which conserve the number of singly occupied AO's. We say that the mixing of CW's of this type defines "covalent" electron delocalization. That is to say, a single electron hop from one AO to another occurs in a manner which does not introduce electron pairing within one and the same AO. An example is given below.

$$\left[\begin{matrix} & x_2 \downarrow & \\ x_1 \uparrow & & x_3 - \end{matrix}\right]_{\Phi_1} \overset{H'}{\longleftrightarrow} \left[\begin{matrix} & x_2 - & \\ x_1 \uparrow & & x_3 \downarrow \end{matrix}\right]_{\Phi_2}$$

c) "Ionic" Delocalization. In this case, H_{ij} connects two CW's which again differ by one occupied spin orbital but which no longer conserve the number of singly occupied AO's. We say that the mixing of CW's of this type defines "ionic" electronic delocalization. That is to say, a single electron hop from one AO to another occurs in a manner which introduces a pairing of two electrons in one and the same orbital.

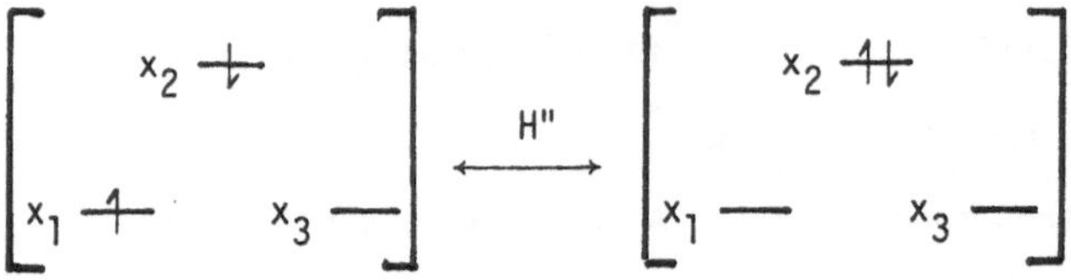

$$\left[\begin{matrix} & \chi_2 \,\uparrow & \\ \chi_1 \,\uparrow & & \chi_3 \,- \end{matrix}\right] \xleftrightarrow{H''} \left[\begin{matrix} & \chi_2 \,\uparrow\downarrow & \\ \chi_1 \,- & & \chi_3 \,- \end{matrix}\right]$$

Φ_1 $\quad$ X_2

d) Bielectronic "Ionic" Delocalization. In this case, H_{ij} connects two CW's which differ by two occupied spin orbitals. An example is given below:

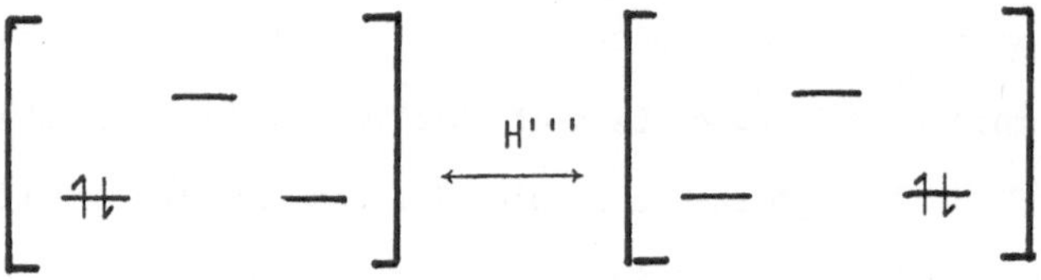

$$\left[\begin{matrix} & - & \\ \uparrow\downarrow & & - \end{matrix}\right] \xleftrightarrow{H'''} \left[\begin{matrix} & - & \\ - & & \uparrow\downarrow \end{matrix}\right]$$

X_1 $\quad$ X_3

It must be pointed out that electron delocalization as defined by H', H'', and H''' is of the interatomic charge transfer (CT) type since the AO's belong to different centers. Analogous definitions for electron delocalization of the intraatomic local excitation (LE) type are obvious.

Two types of CT electron delocalization, namely, "ionic" and "covalent" delocalization, are of particular interest because they are involved in problems of diverse nature. Thus, in the space below, we discuss how "ionic" delocalization is reproduced at various levels of MO and VB theory and how the balance of "ionic" and "covalent" delocalization is "seen" by various theoretical methods. We begin with an examination of "ionic" delocalization using a specific example, the interaction of the HL and "ionic" VB CW's, Φ_2 and Φ_3, respectively, shown in Table 2. The three key measures of interaction, or, equivalently,one-electron "ionic" delocalization computed at various levels of VB theory are presented in Table 4. The following trends are noteworthy:

a) At the HVB level, neglect of differential overlap as well as "classical" coulomb effects, i.e., neglect of exchange and coulomb correlation, produces unrestrained "ionic" delocalization, i.e., mixing of Φ_2 and Φ_3,which is artificially greater than optimal as a result of an artificially small ΔE and a zero CW overlap integral, S_{23}.

b) At the level of EHVB level, neglect of "classical" coulomb effects, i.e., neglect of coulomb correlation, produces unrestrained delocalization as a result of an artifically small ΔE.

c) At the NDO-VB level, neglect of differential overlap, i.e., partial neglect of coulomb correlation and total neglect of exchange correlation, produces unrestrained delocalization as a result of S_{23} being set equal to zero. This discussion reemphasizes the fact that all approximate VB and MO methods produce artificially overdelocalized eigenstates of high ionicity through an incorrect mixing of "covalent" annd "ionic" CW's.

Next, let us compare "ionic" and "covalent" delocalization by computing the energy depression of Φ_1 via its interaction with Φ_2 or X_2, according to equations (100 - 103) at various levels of theory. For simplicity, we assume the following:

$$\varepsilon_1 = \varepsilon_2 = \varepsilon_3 = \varepsilon \quad (105)$$

$$s_{12} = s_{13} = s_{23} = s \quad (106)$$

$$\beta_{12} = \beta_{13} = \beta_{23} = \beta \quad (107)$$

$$s^2 = 0 \quad (108)$$

The results are given in Table 5 and the following trends are noteworthy insofar as the VB expressions are concerned:

a) If we neglect the bielectronic terms, the interaction matrix element, H_{ij}, favors "ionic" or "covalent" delocalization depending on the value of s. Specifically, as s increases, the interaction matrix element, H_{ij}, increasingly favors "covalent" delocalization. In addition, the CW overlap integral, S_{ij}, operates in favor of "covalent" delocalization as $S''>S'$ for all values of s. Accordingly, the critical interaction quantity $H_{ij} - E_i S_{ij}$ favors "covalent" delocalization.

b) The energy separation of the interacting CW's clearly favors "covalent" over "ionic" delocalization. In the former case, the interaction is first order in energy as the interacting CW's are degenerate, while, in the latter case, it becomes second order in energy as a large energy gap separates the interacting CW's. In general, the energy difference, ΔE, of the interacting CW's is much smaller in the case of "covalent" delocalization than in the case of "ionic" delocalization because of two reasons:

1. In the case of "ionic" delocalization, a single electron hop results in a simultaneous occupation of an AO by two electrons. This brings into play severe interelectronic repulsion which destabilizes the closed shell relative to the open shell CW.

2. In the case of "ionic" delocalization, the same single electron hop which results in the pairing of two electrons within an AO reduces the number of singlet coupled unpaired electrons and, as a result, eliminates overlap attraction. Once again, the closed shell CW is destabilized relative to the open shell CW.

The following conclusion becomes apparent: <u>"covalent" delocalization is favored over "ionic" delocalization on account of coulomb and exchange correlation of electronic motions.</u>

Table 5. Energy Expressions for "Ionic" and "Covalent" Delocalization at Various Levels of Theory.

"Covalent" Delocalization

	VB	NDO-VB	EHVB	HVB
H'	$\beta(1 + 2s) + (\varepsilon + V)s + \langle 11\|23\rangle + \langle 13\|21\rangle$	β	$\beta(1 + 2s) + \varepsilon s$	β
S'	s	0	s	0
$\Delta E'$	0	0	0	0
$\delta E'$	equation (102)	equation (102)	equation (102)	equation (102)

"Ionic" Delocalization

	VB	NDO-VB	EHVB	HVB
H''	$\sqrt{2}[\beta + (\varepsilon + V)s + \langle 22\|12\rangle]$	$\sqrt{2}\,\beta$	$\sqrt{2}\,(\beta + \varepsilon s)$	$\sqrt{2}\,\beta$
S''	$\sqrt{2}\,s$	0	$\sqrt{2}\,s$	0
$\Delta E''$	$J_{22} - J_{12} - 2\beta s - K_{12}$	$J_{22} - J_{12}$	$-2\beta s$	0
$\delta E''$	equation (100)	equation (100)	equation (100)	equation (100)

Let us now identify the key deficiencies of approximate methods insofar as the description of the balance of "covalent" and "ionic" delocalization is concerned by reference to Table 5:

a) HVB or HMO theory fails catastrophically because it favors "ionic" delocalization due to complete neglect of coulomb and exchange correlation.

b) A comparison of EHVB or EHMO and NDO-VB theory is difficult. The variation of $H_{ij}-E_iS_{ij}$ in EHVB theory parallels that of VB theory while the opposite is true in the case of NDO-VB theory. On the other hand, the NDO-VB energy gap is a better approximation of the VB energy gap than the EHVB energy gap, at least along most of the reaction coordinate describing bond dissociation (_vide supra_). We shall keep this fact in mind for, at some point along the development of the theory, we shall need to make a choice among different brands of qualitative VB theory for the purposes of tackling actual chemical problems. NDO-VB theory has the advantage of greater simplicity, precisely because of the NDO approximation, and it also yields the right form of the total wavefunction at long interfragmental distances. Thus, this type of approximate VB theory will be most suitable for the discussion of transition states and reactivity, in general.

F. Diagrammatic Matrix Elements

The main reason for the "bad reputation" of the VB method among organic chemists whose theoretical interests were first aroused by the success of the Woodward-Hoffmann rules is the fact that the HL version of VB theory does not expose in any simple manner the origin of aromaticity and antiaromaticity. Thus, two equivalent canonical structures can be written for benzene as well as cyclobutadiene:

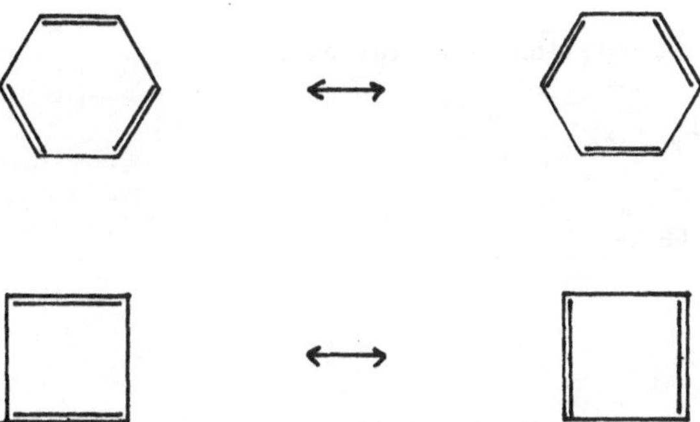

According to "resonance theory", the oversimplified version of VB theory, the two molecules should be equally stabilized. Nonetheless, it is well known that benzene is remarkably stable and cyclobutadiene remarkably unstable.[90] No insights as to why this apparent breakdown of "resonance theory" occurs can be found in the early expositions of formal HL-type VB theory.

In order to be able to understand the origin of this "difficulty", we must take a close look at matrix elements between determinantal wavefunctions. The key ideas can be illustrated by reference to the energy and overlap matrix elements shown below:

$$E = \langle\Phi|\hat{H}|\Phi\rangle \qquad (109)$$

$$S = \langle\Phi|\hat{1}|\Phi\rangle \qquad (110)$$

The AO and fragment notation is as follows:

$$\underset{A}{x_1 \; \uparrow\downarrow} \qquad \underset{B}{\uparrow\downarrow \; x_2}$$

The unnormalized CW is:

$$\Phi = |x_1\bar{x}_1 x_2\bar{x}_2| \qquad (111)$$

The total electrostatic Hamiltonian is:

$$\hat{H} = \sum_i^4 -\frac{1}{2}\nabla_i^2 - \sum_i^4 \left(\frac{Z_A}{r_{iA}} + \frac{Z_B}{r_{iB}}\right) + \frac{1}{r_{12}} + \frac{1}{r_{13}} + \frac{1}{r_{14}} + \frac{1}{r_{23}} + \frac{1}{r_{24}} + \frac{1}{r_{34}} + \frac{Z_A Z_B}{r_{AB}} \qquad (112)$$

$$\hat{H} = \hat{O}' + \hat{O}'' \qquad (113)$$

$$\hat{O}' = \sum_i^4 -\frac{1}{2}\nabla_i^2 - \sum_i^4 \left(\frac{Z_A}{r_{iA}} + \frac{Z_B}{r_{iB}}\right) \qquad (114)$$

$$\hat{O}'' = \sum r_{ij}^{-1} + \frac{Z_A Z_B}{r_{AB}} \qquad (115)$$

Finally, we have:

$$E = \sum_{\hat{P}} (-1)^P \langle x_1(1)\bar{x}_1(2)x_2(3)\bar{x}_2(4)|\hat{O}' + \hat{O}''|\hat{P}x_1(1)\bar{y}_1(2)y_2(3)\bar{x}_2(4)\rangle \qquad (116)$$

Since most experimentalists are unfamiliar with matrix elements over determinantal wavefunctions constructed from nonorthogonal orbitals, we have sought to develop E in Scheme 1 in a way which shows clearly what parts of the operator produce the monoelectronic and bielectronic parts, how different types of permutations bring into play "classical" and "nonclassical" effects, and how each exchange term is broken down to $E'(P_{xy}\ldots)$,or,T_n,and $E''(P_{xy}\ldots)$,or,R_n,parts which are defined in such a way so that T_n contains all terms included in EHMO theory. With regards to the last issue, each exchange term [e.g., $E(P_{13})$] is written in a way so that the terms inside the dotted lines comprise T_n and those outside comprise R_n. The value of E is obtained by summing all terms of Scheme 1.

The total energy of Φ is a sum of four energy terms generated by the four possible electron exchanges which do not produce spin orthogonalities. Thus, $E(P^0)$ represents the energy term generated by zero exchange of the electrons in the diagonal element of the Slater determinant. $E(P_{13})$ represents the energy term generated by exchange of electrons 1 and 3 in the diagonal element of the Slater determinant, and so on. When the fragments are not oppositely charged so as to bring into play coulomb attraction, we can assume that G is zero. Furthermore, we can neglect the R_n components of EX as being mere attenuators of the larger T_n terms. With these approximations, we obtain:

$$E \simeq F + EX' \tag{117}$$

$$E \simeq F + T_1 + T_2 + T_3 \tag{118}$$

Our goal now is to develop a pictorial representation of the "monoelectronic" part of each energy exchange term, T_n, in a way which will facilitate our analysis of the electronic structure of chemical systems. A convenient diagrammatic representation of each term of this type can be obtained if we use the Wolfsberg-Helmholz approximation of the AO "resonance integral", as given

Scheme 1 Analytical Construction of Diagonal Hamiltonian Matrix Element for the 4 electron 2 center system.

		Monoelectronic Terms				Bielectronic and Nuclear Terms						
Resultant Term	Permutation	$0_1'$	$0_2'$	$0_3'$	$0_4'$	$\frac{1}{r_{12}}$	$\frac{1}{r_{13}}$	$\frac{1}{r_{14}}$	$\frac{1}{r_{23}}$	$\frac{1}{r_{24}}$	$\frac{1}{r_{34}}$	
F + G	$\underline{P^o}$	ε_A V_B	ε_A V_B	ε_B V_A	ε_B V_A	J_{AA}	J_{AB}	J_{AB}	J_{AB}	J_{AB}	J_{BB}	"Classical" Terms
						V_{nn}						
T_1 * + R_1	$\underline{P_{13}}$	$-\beta s$	$-\varepsilon_A s^2$ $-V_B s^2$	$-\beta s$	$-\varepsilon_B s^2$ $-V_A s^2$	$-<AB\|AA>s$	$-K_{AB}$	$-<BB\|AB>s$	$-<AB\|AA>s$	$-J_{AB}s^2$	$-<AB\|BB>s$	"Nonclassical" or Exchange Terms
						$-V_{nn}s^2$						
T_2 * + R_2	$\underline{P_{24}}$	$-\varepsilon_A s^2$ $-V_B s^2$	$-\beta s$	$-\varepsilon_B s^2$ $-V_A s^2$	$-\beta s$	$-<AB\|AA>s$	$-J_{AB}s^2$	$-<AA\|AB>s$	$-<AB\|BB>s$	$-K_{AB}$	$-<BB\|AB>s$	
						$-V_{nn}s^2$						
T_3 * + R_3	$\underline{P_{13,24}}$	βs^3	$+\beta s^3$	$+\beta s^3$	$+\beta s^3$	$+Ks^2$	$+Ks^2$	$+Ks^2$	$+Ks^2$	$+Ks^2$	$+Ks^2$	
						$V_{nn}s^4$						

* T_n is contained within the dotted line.

by equation (14) with the assumption that the two fragments A and B are identical. With this in mind, we can now diagram each "monoelectronic" energy exchange term by using the following conventions:

a) A point represents a spin orbital. If two spin orbitals describe the same spatial orbital, the points are connected by a solid line.

b) A wavy line connecting two points represents the corresponding AO overlap integral.

c) A circle about a point represents the one-electron energy of the corresponding spin orbital.

T_1, T_2, and T_3, can be represented in diagrammatic form in the following way:

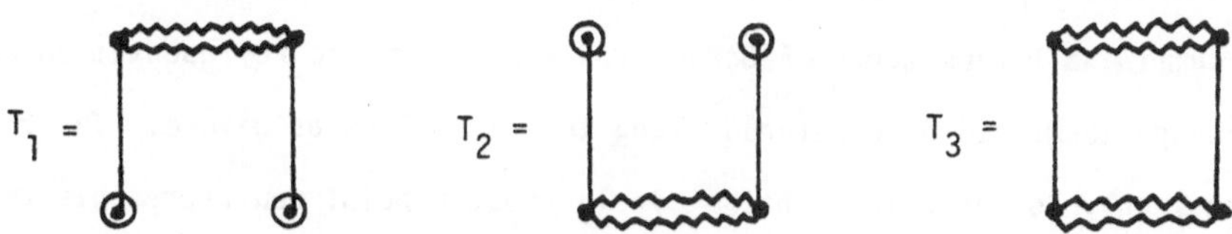

By reference to Scheme 1, we realize that each T_n is comprised of four energy terms, or, as many energy terms as there are occupied spin orbitals. In turn, the number of occupied spin orbitals equals the number of circles and wavy lines in each T_n. We now specify how each diagram is translated into its four constituent energy terms.

a) Each circle is multiplied by the wavy line product.

b) Each wavy line is multiplied by the other wavy line(s), times one circle, times the energy constant K.

The procedure is illustrated for T_1:

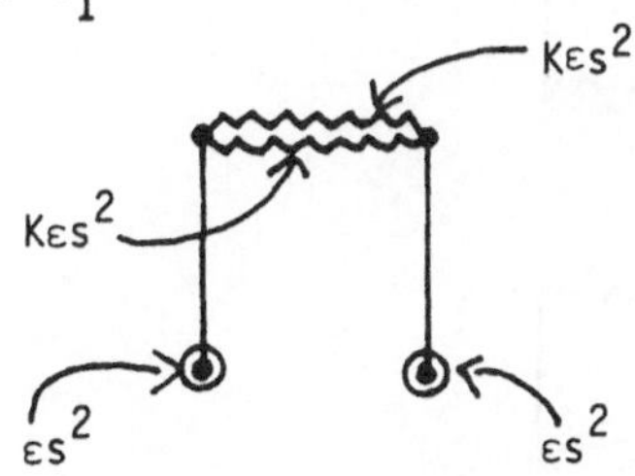

It must be always kept in mind that the pictorial representation of T_n's proposed above is founded on the Wolfsberg-Helmholz approximation of β and it is valid for homonuclear systems. A more complicated recipe for heteronuclear systems can be constructed.

The overlap intergral, S, can be dealt with in a similar manner. In evaluating overlap integrals, the effective operator is unity:

$$S = \langle\Phi|\hat{1}|\Phi\rangle \tag{110}$$

Hence, we have

$$S \simeq \sum_{\hat{P}} (-1)^P \langle x_1(1)\bar{x}_1(2)x_2(3)\bar{x}_2(4)|\hat{1}|x_1(1)\bar{x}_1(2)x_2(3)\bar{x}_2(4)\rangle \tag{119}$$

Once again, S is developed pictorially in Scheme 2. The value of S is obtained by summing all four terms of Scheme 2. A diagrammatic representation of the exchange terms can be developed along the same lines as before. The conventions change only to the extent that a circle about a point now represents unity. Recognizing that the operator generates one term per electron exchange, rather than four as in the case of E, we can write:

$$U_1 = \quad = s^2$$

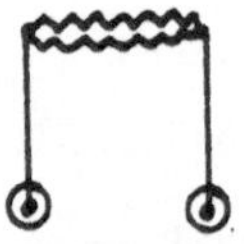

$$U_2 = \quad = s^2$$

$$U_3 = \quad = s^4$$

Scheme 2 Analytical Construction of Diagonal Overlap Matrix Element for the 4 electron - 2 center System.

Term	Permutation	Result	
"Classical"	P^{o}	1	"Classical" Term
U_1	P_{13}	$-s^2$	"Non-Classical" or Exchange Terms
U_2	P_{24}	$-s^2$	
U_3	$P_{13,24}$	s^4	

Scheme 3 Diagrammatic Matrix Elements for the 4 electron - 2 center System[a]

	P^{o}	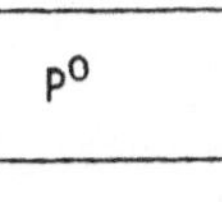	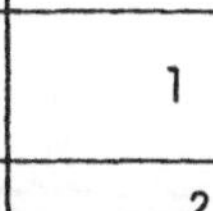	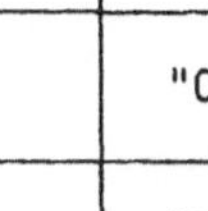
E =	F	1	1	1
S =	1	1	1	1

(a) Energy expressions for the T_n are shown in Appendix 1.

The recipe for the generation of a given overlap exchange term from the appropriate diagram amounts to simply forming the direct product of circles and wavy lines. The procedure is exemplified by reference to the overlap exchange term U_1:

Wavy line x wavy line x 1 x 1 = s^2

The final results are presented in tabular form in Scheme 3. Recall that, in general, the total energy expression of a VB CW is a sum of F+G, X', and X''. However, G can usually be neglected, especially if the fragments are neutral, and X'' simply attenuates the effect of X'. Hence, a representation of E and S such as that shown in Scheme 3 is a good description of E and S for most systems of interest. Scheme 3 is said to be the diagrammatic representation of the matrix elements E and S on the basis of the stated approximations. In summary, the diagrammatical representation of matrix elements is the key to a pictorial understanding of chemical bonding. We shall have ample opportunity to appreciate its value when we tackle much more complex matrix elements.

Before we proceed any further, we note that we can render diagrammatic matix elements more concise in two important ways:

a) An important consequence of the conventions specified above is that a single diagram represents the T_n term of H_{ij} as well as the corresponding U_n term of S_{ij}. This has the advantage that a diagrammatic representation of H_{ij} can be "read" to produce the corresponding S_{ij}. Accordingly, in applying these techniques to actual problems, we shall omit the representation of S_{ij} as it is directly derivable from that of H_{ij}.

b) A more compact diagrammatic representation can be obtained by replacing the two spin orbitals x_i^{α} and x_i^{β} by the spatial orbital x_i. Accordingly, the T_n's and U_n's can be redrawn in the manner illustrated below.

T_1 =

U_1 =

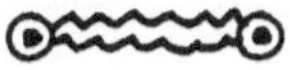

In most applications, a diagrammatic representation over simple AO's is consistently possible. However, there exist cases in which a "compromise", or, "mixed", diagrammatic representation becomes necessary. Henceforth, we shall be using the compact representation to the extent possible.

The diagrammatic treatment of "monoelectronic" energy and overlap terms outlined above makes pictorially evident that an important aspect of all energy and overlap matrix elements is their dependence on the product of AO overlap integrals. We distinguish two extreme situations:

a) All overlap integrals have an even exponent (i.e., an even number of wavy lines connect two orbitals).

b) At least one overlap integral has an odd exponent (i.e., an odd number of wavy lines connect two orbitals).

"Monoelectronic" energy and overlap terms of the first type will be referred to as even and those of the second type as odd. The same nomenclature will be used for the corresponding "monoelectronic" energy and overlap diagrams. Illustrative examples of even and odd diagrams are given in Figure 2. Naturally, even terms are invariant to sign reversal of the AO overlap integrals while odd terms are dependent upon such a sign reversal. It follows that whenever the energy and overlap matrices generated by the variational treatment of a trial VB wavefunction contain odd terms, the final energies of the VB eigenstates will depend on the choice of signs of the AO overlap integrals which have odd exponents. As we shall presently see, this is the mathematical basis of streoselectivity at the level of VB theory.

Number of Centers	Even Terms	Odd Terms
2	 a b a b	
3	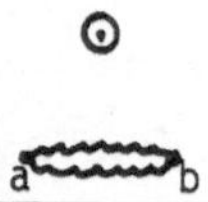 a b	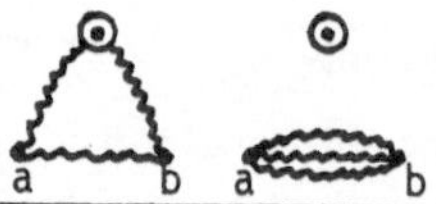 a b a b
4	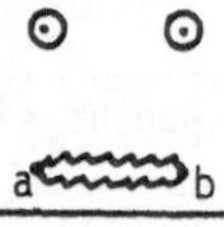 a b	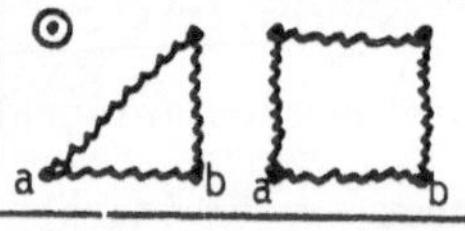 a b a b

Figure 2: Illustrative samples of even and odd energy and overlap diagrammatic matrix elements (even and odd determined with respect to interaction of centers a and b).

G. The VB Theory of Aromaticity

In HMO theory, the basis set is made up of AO's, and the energy and overlap matrix elements are defined as follows:[91]

$$\langle \chi_t | \hat{H} | \chi_t \rangle = \alpha_t \qquad (120)$$

$$\langle \chi_t | \hat{H} | \chi_u \rangle = \beta_{tu} \qquad (121)$$

$$\langle \chi_t | \chi_t \rangle = 1 \qquad (122)$$

$$\langle \chi_t | \chi_u \rangle = s_{tu} \qquad (123)$$

$\beta_{tu} \alpha\, s_{tu}$ [See equation (14)]

It is evident that only the off diagonal matrix elements depend on overlap and, in particular, on an odd power of the AO overlap integral, s_{tu}. A system wherein the AO's overlap in a noncyclic manner does not exhibit energy dependence on the sign(s) of the AO overlap integrals but a system wherein the AO's overlap in a cyclic manner does. For example, in a system comprised of three AO's which overlap in a cyclic manner, a stereochemical distinction between an aromatic and an antiaromatic geometry is possible as the total energy expression depends on the signs of the three s_{tu}'s. Different assignments generate Hückel or Möbius aromatic or antiaromatic arrays depending on the number of electrons as indicated below.

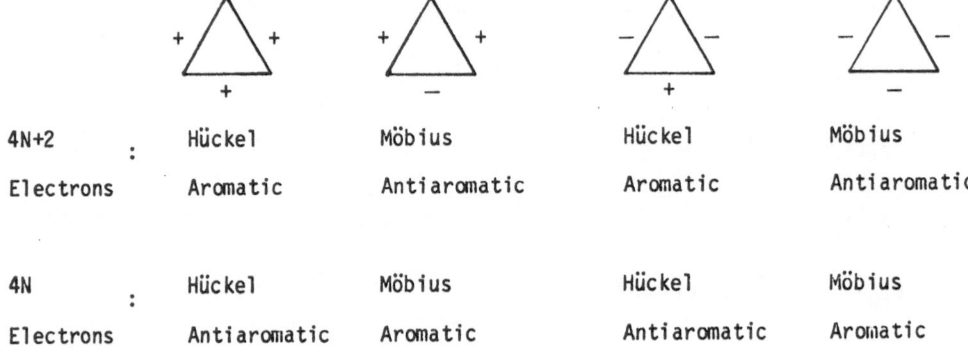

4N+2 Electrons	:	Hückel Aromatic	Möbius Antiaromatic	Hückel Aromatic	Möbius Antiaromatic
4N Electrons	:	Hückel Antiaromatic	Möbius Aromatic	Hückel Antiaromatic	Möbius Aromatic

A Hückel cyclic AO array is characterized by a positive product of AO overlap integrals and a Möbius cyclic AO array by a negative one.

In VB theory, the situation is considerably different because the basis set is now made up of VB CW's and both diagonal as well as off diagonal matrix elements are functions of AO overlap. The AO overlap dependence of the VB energy matrix elements given by equations (58), (65), and (70) is made evident from considerations of the AO overlap dependence of the constituent terms. This is as follows:

$F_{aa} + G_{aa}$: Even

EX_{aa} : Even or Odd

H^{o}_{ab} : Even or Odd

EX_{ab} : Even or Odd

The parity of the EX_{aa}, H°_{ab}, and EX_{ab} terms is determined by the type of spin orbital occupancy.

We can now use the system of equations (49)-(52) in order to cast the VB matrix elements into the following forms where M and M' are constants.

$$H_{ii} = M(F + G) \pm (\sum_n T_n^{odd} + \sum_n R_n^{odd} + \sum_m T_m^{even} + \sum_m R_m^{even}) \quad (124)$$

$$S_{ii} = M \cdot 1 \pm (\sum_n U_n^{odd} + \sum_m U_m^{even}) \quad (125)$$

If Φ_i and Φ_j have common X_a's:

$$H_{ij} = M'(F + G) \pm (\sum_n T_n^{odd} + \sum_n R_n^{odd} + \sum_m T_m^{even} + \sum_m R_m^{even}) \quad (126)$$

$$S_{ij} = M' \cdot 1 \pm (\sum_n U^{odd} + \sum_m U^{even}) \quad (127)$$

If Φ_i and Φ_j have no common X_a's:

$$H_{ij} = \pm (\sum_n T_n^{odd} + \sum_n R_n^{odd} + \sum_m T_m^{even} + \sum_m R_m^{even}) \quad (128)$$

$$S_{ij} = \pm (\sum_n U_n^{odd} + \sum_m U_m^{even}) \quad (129)$$

Neglecting the "attenuating" "bielectronic" exchange terms R_n and recalling that each T_n or U_n corresponds to a specific diagram as prescribed in the previous section we make note of two important points:

a) The sign of an AO overlap integral determines whether a T_n^{odd} diagram "adds" or "subtracts" in the expansions (124) to (129).

b) Since the first term of the right hand side of each of equations (124) to (127) is by far the larger of the two, the signs of the corresponding matrix elements will be independent of the signs of the AO overlap integrals. On the other hand, their magnitudes will be dependent upon the signs of the AO overlap integrals.

c) Both magnitude and sign of each of the matrix elements given by equations (128) and (129) will be dependent upon the signs of the AO overlap integrals.

At this point, we note that a <u>given choice of AO overlap integral signs corresponds to a distinct and realizable stereochemical arrangement of the constituent atoms of a system</u>. Furthermore, a sign reversal of a given AO overlap integral is tantamount to a transition from a system of one symmetry to another with a different symmetry. With this in mind, we conclude that symmetry expresses itself in both diagonal and off diagonal matrix elements over VB CW's insofar as it exerts a controlling influence on the magnitudes of the diagonal and the magnitudes as well as signs of off diagonal terms. At the VB level, the diagrammatic representation of matrix elements leads to a simple, yet detailed, mathematical, yet pictorial, understanding of the electronic basis of stereoselection.

We are now finished with the formulation of definitions and concepts which will be needed in tackling a variety of problems with the VB method. In this part, we shall examine the electronic basis of stereoselection, i.e., aromaticity and antiaromaticity, from the vantage point of a number of rigorous and approximate VB theories. Our intent is to illustrate the "how to do it" aspect of different brands of VB theory rather than to shock with the novelty of the conclusions. The "surprise element" will creep into most, if not all, of the following papers. These papers, however, can only be understood after the reader has gained some familiarity with the inner workings of VB theory and its extensions.

We shall illustrate our approach by reference to three prototypical systems:

1. A two electron-three center system, denoted by 2e-3c, e.g., cyclic H_3^+, pi cyclopropenyl cation, etc.

2. A four electron-three center system, denoted by 4e-3c, e.g., cyclic H_3^-, pi cyclopropenyl anion, etc.

3. A four electron-four center system, denoted by 4e-4c, e.g., square H_4, pi cyclobutadiene, etc.

In each case, we shall seek to understand why "aromatic is better than antiaromatic" at the levels of HL, HVB, and NDO-VB theories.

H. Heitler-London Theory of Stereoselection

One important goal of this work is the development of qualitative VB theories which can be profitably applied to diverse chemical problems. For reasons that will become clear in this and future works, we define the following brands of Approximate HL (AHL) theory:

a) AHL theory in which we set G and R_n equal to zero and we truncate each T_n and U_n so that they include terms which are zero, first, or, second order in AO overlap. This brand of HL theory will be denoted by AHL*.

b) AHL theory in which we set G and R_n equal to zero but we do not truncate T_n and U_n. This brand of HL theory will be denoted by AHL°.

c) AHL theory in which we set only R_n equal to zero and we truncate T_n and U_n as indicated in (a). This brand of HL theory will be denoted by AHL‡.

d) AHL theory in which we set R_n equal to zero but no further approximations are made. This will be simply referred to as AHL theory.

A qualitative understanding of how and why chemical stereoselection arises can be obtained through AHL° theory. At this level of theory the matrix element defined by equations (124) to (129) take the following forms:

$$H_{ii} = (M)F + X_i' \tag{130}$$

$$S_{ii} = (M)1 + SX_i \tag{131}$$

$$H_{ij} = (M')F + X_{ij}' \tag{132}$$

$$S_{ij} = (M')1 + SX_{ij} \tag{133}$$

$$H_{ij} = X_{ij}' \qquad (\Phi_i \text{ and } \Phi_j \text{ have no common } X_a\text{'s}) \tag{134}$$

$$S_{ij} = SX_{ij} \qquad (\text{" " " " " " "}) \tag{135}$$

In a qualitative sense, a diagrammatic representation of CW matrix elements is equivalent to a solution of the stereoselection problem.

Next, we make a careful distinction between Hückel or Möbius AO arrays and Hückel or Möbius CW arrays, and note that there may or may not be a one to one correspondence between the two entities. Thus, for example, we shall see that in 3-center annulenes, a Hückel AO array generates a Hückel HL CW array and a Möbius AO array generates a Möbius HL CW array when the total number of electrons is 2. On the other hand, a Hückel AO array generates a Möbius HL CW array and a Möbius AO array generates a Hückel HL CW array when the total number of electrons is 4. Since the lowest eigenvalue of a given cyclic CW array corresponds to the ground state of the system in question and since the lowest eigenvalue of a Hückel CW array is more negative than that of a Möbius CW array we shall be able to redefine ground state aromaticity by focussing attention on the type of CW interaction promoted by a given system.

Finally, we convene that all diagrammatic representations of matrix elements are written assuming that all AO overlap integrals are positive. In the case of cyclic systems, the convention is that diagrammatic matrix elements are drawn for Hückel AO arrays.

2e-3c Systems

The three HL CW's required for the description of a cyclic 2e-3c system are shown below.

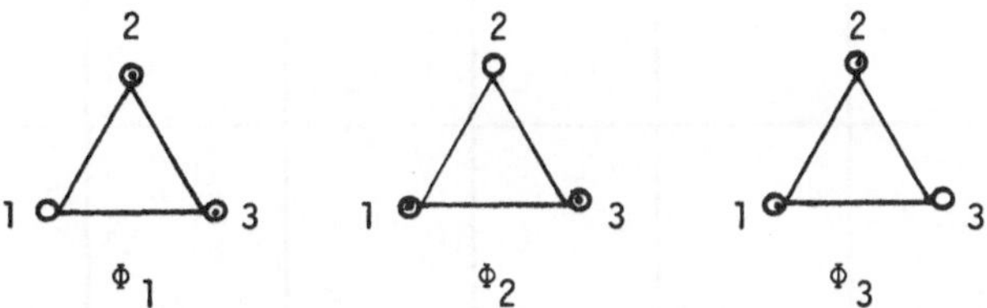

The various matrix elements are shown in diagrammatic form in Scheme 4. The following are noteworthy trends:

a) Each T_n is made up of two terms since there are two electrons and, consequently, two occupied spin orbitals. In other words, each T_n must contain two wavy lines or one wavy line and one circle.

Scheme 4 Diagrammatic Matrix Elements for the 2 electron - 3 center System [a,b]

1 2 3	F	T_1'	T_1''	T_1'''	T_2'	T_3'	T_2''	T_3''	T_2'''	T_3'''
H_{11}	2	2								
H_{22}	2		2							
H_{33}	2			2						
H_{12}					2	2				
H_{13}							2	2		
H_{23}									2	2

(a) Energy expressions for the T_n are given in Appendix 1.

(b) $F = 2\varepsilon$

b) Since there are three equivalent CW's, there are three equivalent T_n's for every value of n denoted by one, two, and three primes on T. For the problem at hand, we have the following relationships:

$$T_1 = T_1' = T_1'' = T_1''' \tag{136}$$

$$T_2 = T_2' = T_2'' = T_2''' \tag{137}$$

$$T_3 = T_3' = T_3'' = T_3''' \tag{138}$$

c) The largest exchange term is always the one with the fewest wavy lines. In this case, it is the "semiclassical" term T_2. The energy expressions for the T_n's are given in Appendix 1.

d) T_1', T_1'', and T_1''' are even diagrams while T_2', T_2'', T_2''', T_3', T_3'', and T_3''' are odd diagrams. Reversing the sign of an AO overlap integral will have no effect on the former three but it will affect some but not all of the latter six.

The diagrammatic representation of the matrix elements given in Scheme 4 is pertinent to a Hückel AO array. The diagrammatic representation of the matrix elements pertinent to a Möbius AO array can be obtained by simply reversing the sign of an AO overlap integral and noting its effect on every single T_n. Inspection of Scheme 4 reveals that reversal of the sign of s_{13} will have the following consequences:

a) The signs of the even T_1', T_1'', T_1''' <u>and</u> the odd T_2' T_2'', and T_3''' will remain unaffected, while the signs of the odd T_2''', T_3', and T_3'' will be reversed. Accordingly, the matrix elements of the Hückel and Möbius AO arrays become as follows:

	Hückel	Möbius
$H_{11}=H_{22}=H_{33}$	$2F + 2T_1$	$2F + 2T_1$
H_{12}	$2T_2 + 2T_3$	$2T_2 - 2T_3$
H_{13}	$2T_2 + 2T_3$	$2T_2 - 2T_3$
H_{23}	$2T_2 + 2T_3$	$-2T_2 + 2T_3$

Two things become very evident:

a) The diagonal matrix elements, being even functions of AO overlap, are invariant with regards to sign or magnitude.

b) The off diagonal matrix elements, being odd functions of AO overlap, differ with regards to sign and magnitude depending on whether the AO array is of the Hückel or Möbius type. In the former case, the three CW's are connected by three H_{ij}'s which are preceded by positive signs, while, in the latter case, H_{12} and H_{13} are preceded by positive and H_{23} by negative signs. Thus, the Hückel AO array gives birth to a Hückel CW array and the Möbius AO array to a Möbius CW array. The sign interrelationships of the three interaction matrix elements are shown schematically below.

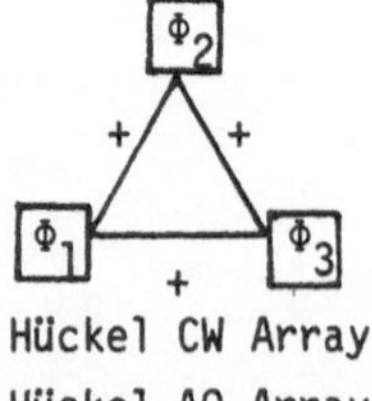

Hückel CW Array
Hückel AO Array

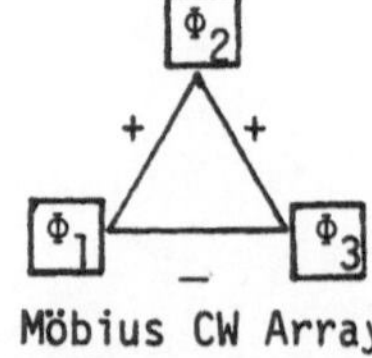

Möbius CW Array
Möbius AO Array

c) As demanded by the symmetry of the problem, all three interaction matrix elements are equal in each system. But, the interaction matrix elements of the Hückel AO array are larger in absolute magnitude than those of the Möbius AO array.

The above analysis suggests that aromaticity and antiaromaticity in 2e-3c annulenes is expressed via the signs and magnitudes of the interaction matrix elements. If overlap were neglected, the energies of the eigenstates of the Hückel and Möbius AO arrays could be written down from mere knowledge of HMO theory as indicated in Figure 3. This is so because the HMO theoretical approximations lead to the following HL expressions:

$$F = 2\alpha \tag{139}$$

$$T_1 = 0 \tag{140}$$

$$T_2 = \beta \tag{141}$$

$$T_3 = 0 \tag{142}$$

As a result, we must diagonalize the energy matrices (over normalized Φ_i's) implied by the diagrams shown below:

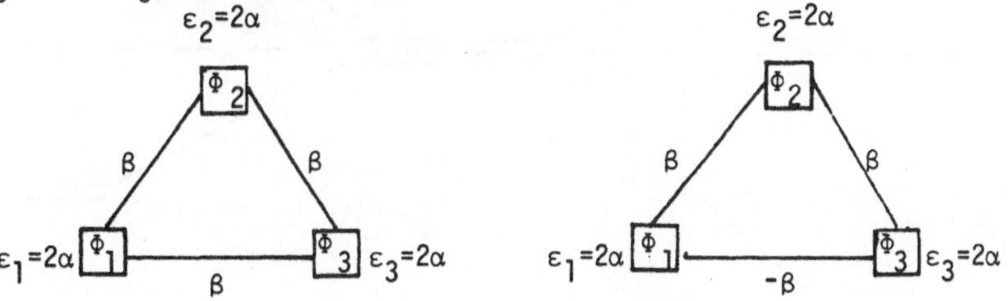

However, replacement of Φ_i by 2p and 2α by α produces the HMO energy matrices for the treatment of Hückel and Möbius cyclopropenium. Hence, the sought after eigenvalues are the same as those obtained in the HMO treatment of Hückel and

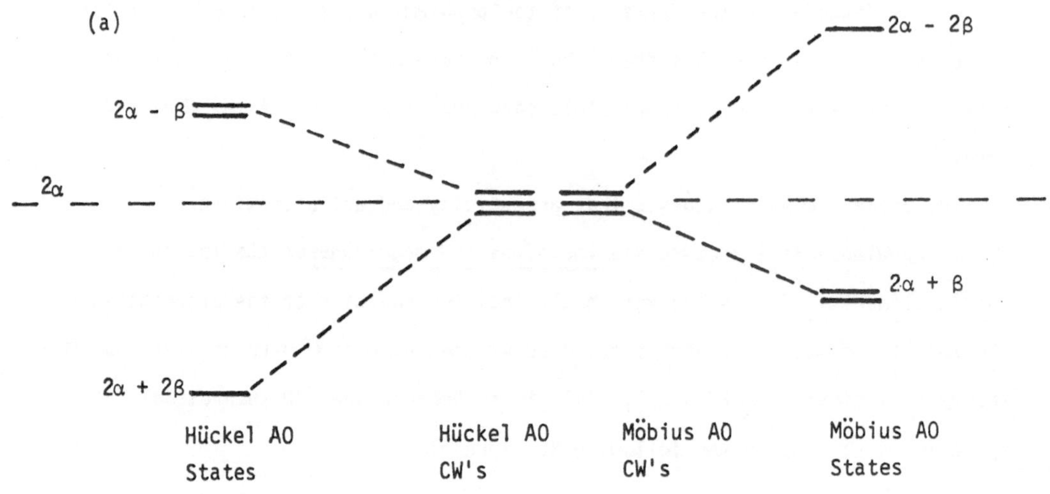

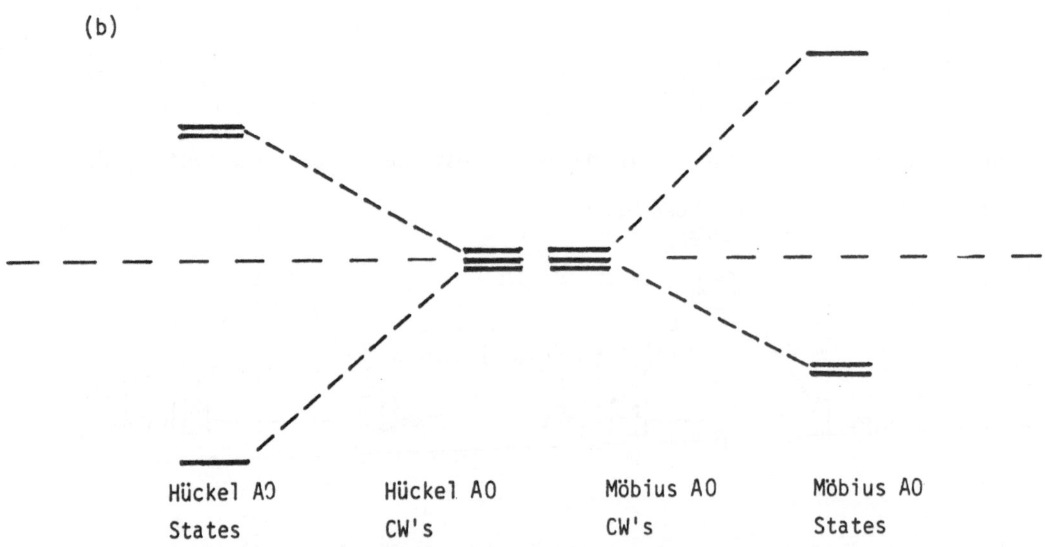

Figure 3: Eigenstates and associated energies for the 2e-3c system:

a) Computations using the familiar HMO (with neglect of overlap) approximations.

b) AHL° interaction diagram.

Möbius cyclopropenium with the only exception that α is replaced by 2α. If overlap is not neglected, the signs as well as the magnitudes of the interaction matrix elements become conveyors of aromaticity and antiaromaticity. This occurs because the off diagonal, but not the diagonal, matrix elements are odd functions of overlap as made evident by the diagrammatic representation of the matrix elements.

4e - 3c Systems

The analysis proceeds along the same lines as in the previous section. The necessary HL CW's are shown below and the various matrix elements are shown in diagrammatic form in Scheme 5.

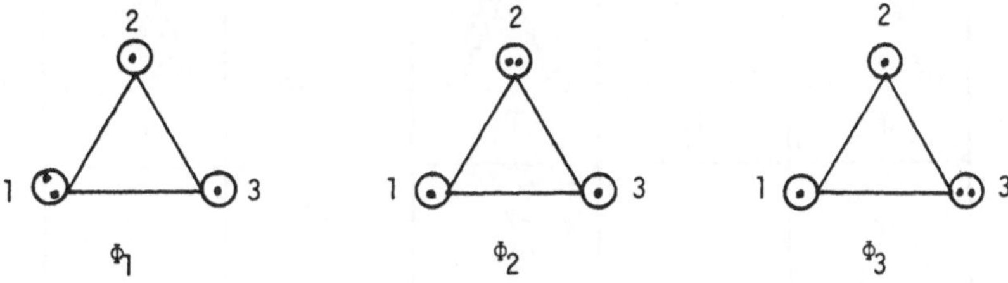

Each T_n is made up of four terms as there are four electrons. This forces us to use spin orbital notation in the case of T_1. For space conservation, we show only one of the three equivalent diagonal and only one of the three equivalent off diagonal elements.

The following things now become evident:

a) The diagonal matrix elements, being odd functions of AO overlap, exhibit stereochemical dependence favoring the Möbius AO system on account of magnitude.

b) The off-diagonal matrix elements, being also odd functions of AO overlap, exhibit a similar stereochemical dependence favoring the Möbius AO system on account of sign and magnitude. The sign effect is due to the fact

Scheme 5 Diagrammatic Matrix Elements for 4 electron - 3 center System [a,b]

	F	 T_1	 T_2	 T_3	 T_4	 T_5
H_{22}	2	2	-2	-2	-4	4

	 T_6	 T_7	 T_8	 T_9	 T_{10}	 T_{11}
H_{12}	-2	4	-2	-2	2	0

(a) Energy expressions for the T_n are given in Appendix 1.

(b) $F = 4\varepsilon$

that in the Hückel AO system, the three CW's are connected by three matrix elements preceded by a negative sign and, thus, they define a Möbius CW array. By contrast, in the Möbius AO system, two CW's are connected by matrix elements preceded by a negative sign and one by a matrix element preceded by a positive sign, and, thus, these three CW's define a Hückel CW array.

The sign interrelationship of the three interaction matrix elements in the Hückel and Möbius arrays is shown schematically below:

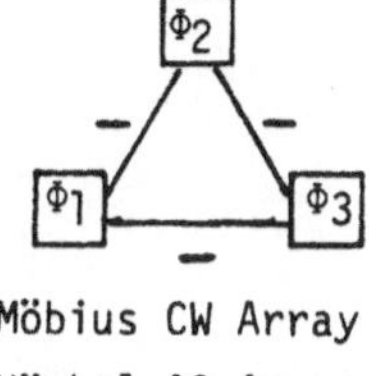

Möbius CW Array
Hückel AO Array

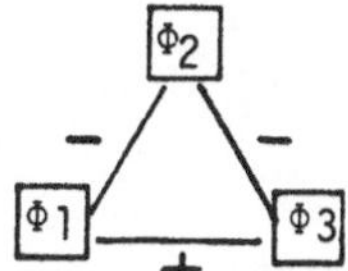

Hückel CW Array
Möbius AO Array

The situation is exactly the reverse of the one encountered in the case of the 2e-3c system where the Hückel AO system generates a Hückel CW array while the Möbius AO system generates a Möbius CW array. Thus, Möbius eigenstates are now connected to the Hückel AO array and Hückel eigenstates are connected to the Möbius AO array.

As in the case of the 2e-3c system, the above analysis suggests again that aromaticity and antiaromaticity in such 4e-3c annulenes is expressed via <u>the signs and magnitudes</u> of the interaction matrix elements, which, in turn, are determined by the signs of the AO overlap integrals. If overlap were neglected we would obtain the eigenstate energies shown in Figure 4a. If overlap is not neglected, the signs of the interaction matrix elements as well as the magnitudes of all matrix elements become conveyors of aromaticity or

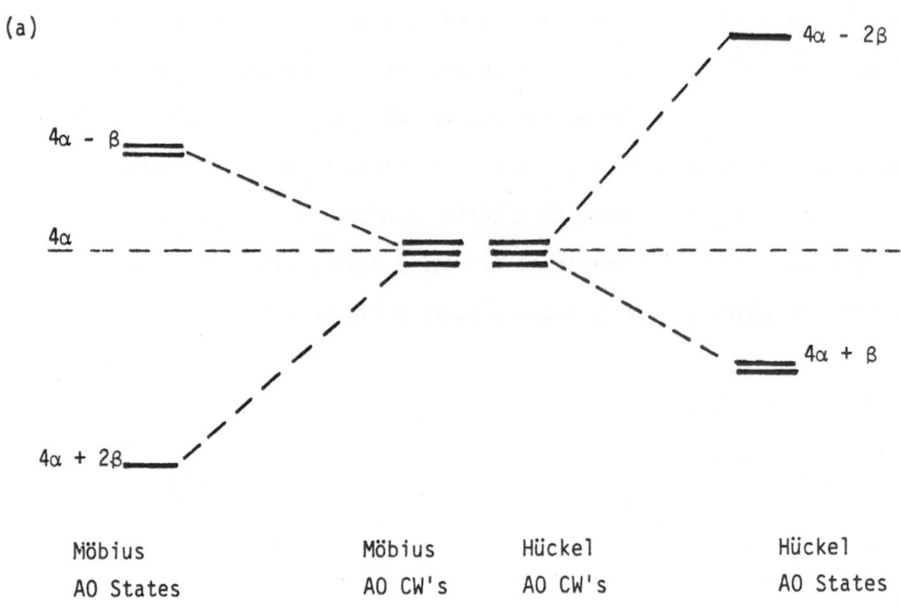

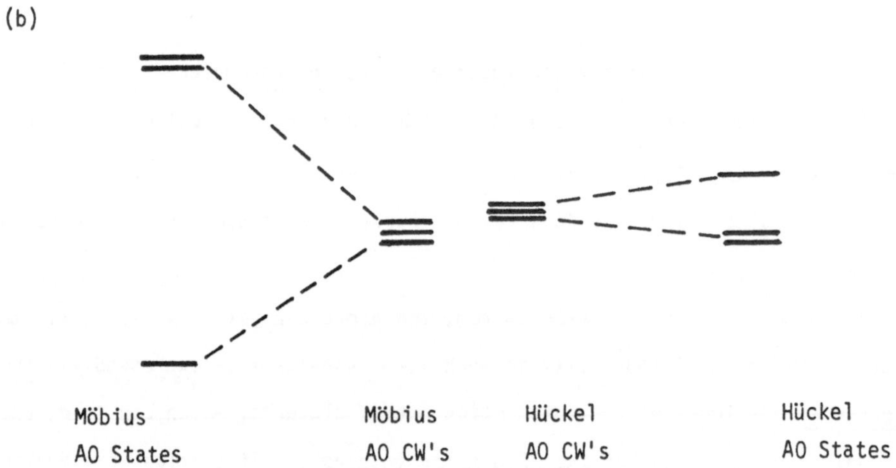

Figure 4: Eigenstates and associated energy for the 4e-3c system:

a) Computations using the familiar HMO (with neglect of overlap) approximations.

b) AHL° interaction diagram.

antiaromaticity.

4e - 4c Systems

The two HL CW's required for the description of a 4e-4c system are shown below. The various matrix elements are shown in diagrammatic form in Scheme 6.

Φ_1 and Φ_2 are the two Kekule CW's (Kekule structures) of the 4e-4c system. A mere inspection of the diagrammatic energy matrix elements reveals that all are odd functions of overlap and that a single AO overlap integral sign reversal lowers the energy of each one of them. Note that the distinction between the two systems comes via terms which are proportional to the third and fourth power of the AO overlap integral. The resulting eigenstates and their associated energies are shown in Figure 5 which makes plain that aromaticity and antiaromaticity in 4e-4c annulenes are less stringently imposed than in 4e-3c annulenes.

The discussion presented above was based upon consideration of diagrammatic energy matrix elements which are not normalized. For example in the case of the 4e-4c system, we considered the unnormalized, H_{11}, rather than the normalized, H_{11}/S_{11}, form of the $<\Phi_1|\hat{H}|\Phi_1>$ matrix element. Now, since H_{11} and S_{11} have the same form, as the coefficients of T_n and U_n are identical, a reversal of the sign of one AO overlap integral decreases or increases, in absolute magnitude, both H_{11} and S_{11}. Since S_{11} acts as a divisor of H_{11}, it is now unclear whether a AO overlap sign reversal which increases H_{11} in absolute magnitude will also

Scheme 6 Diagrammatic Matrix Elements for the 4 electron-4 center System[a,b]

	F	 T_1	 T_2	 T_3	 T_4	 T_5	 T_6
H_{11}	4	-2	-2	4	4	-2	-2
H_{22}	4	-2	-2	-2	-2	4	4
H_{12}	2	-4	-4	2	2	2	2

	 T_7	 T_8	 T_9	 T_{10}	 T_{11}	 T_{12}	T_{13}
H_{11}	-4	-4	-4	-4	4	4	-4
H_{22}	-4	-4	-4	-4	4	4	-4
H_{12}	-2	-2	-2	-2	2	2	-8

a) Energy expressions for representative T_n are shown in Appendix 1.

b) $F = 4\varepsilon$

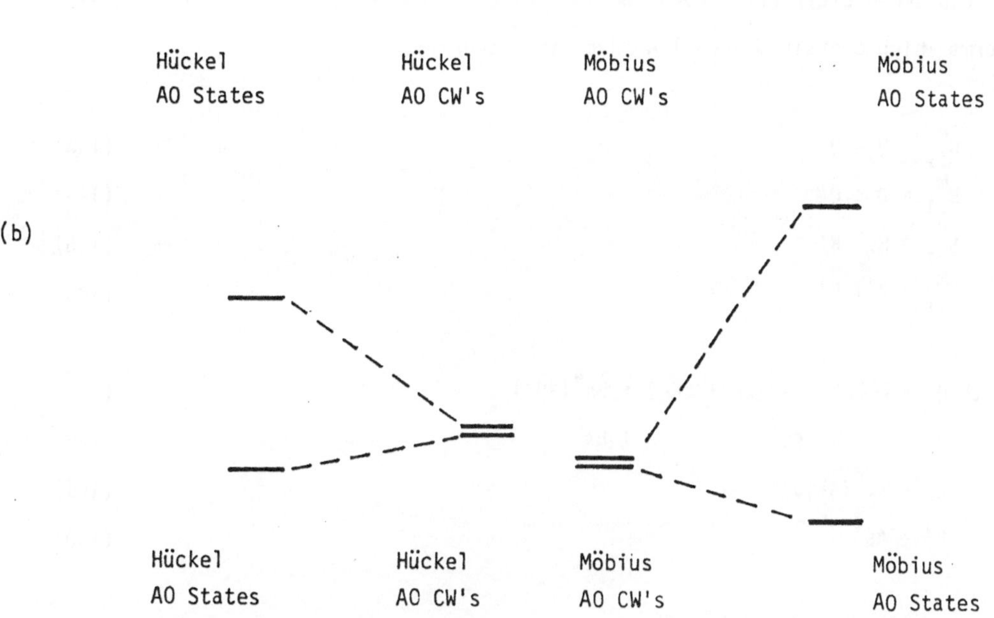

Figure 5: Eigenstates and associated energy for the 4e-4c system:

a) Computations using the familiar HMO (with neglect of overlap) approximations.

b) AHL° interaction diagram.

increase H_{11}/S_{11} in absolute magnitude. If our HL analysis of stereoselection based upon exclusive consideration of the unnormalized H_{ij}'s is to be convincing, we must show that the rate of change of H_{11} dominates that of S_{11} as a transition is made from a Hückel to a Möbius AO system, or vice versa.

From the diagrammatic representation of matrix elements shown in Scheme 6, we can immediately write down the following expressions, where we have neglected terms which contain diagonal AO overlap integrals:

$$H_{11}^{H} = Q - Q' \qquad (143)$$

$$H_{11}^{M} = Q + Q' \qquad (144)$$

$$S_{11}^{H} = R - R' \qquad (145)$$

$$S_{11}^{M} = R + R' \qquad (146)$$

with $Q = 4(4\varepsilon) + 4s^2(2\varepsilon + 2K\varepsilon) + 8s^4(4K\varepsilon)$ (147)

$R = 4 + 4s^2 + 8s^4$ (148)

$Q' = 4s^4(4K\varepsilon)$ (149)

$R' = 4s^4$ (150)

We have shown that $|H_{11}^{M}| > |H_{11}^{H}|$. Now we must show that

$$\left|\frac{H_{11}^{M}}{S_{11}^{M}}\right| > \left|\frac{H_{11}^{H}}{S_{11}^{H}}\right| \qquad (151)$$

or

$$\left|\frac{H_{11}^{M}}{H_{11}^{H}}\right| > \left|\frac{S_{11}^{M}}{S_{11}^{H}}\right| \qquad (152)$$

By dividing H_{11}^{M} and H_{11}^{H} by $4K\varepsilon$, we recast the last inequality in the form

$$\frac{N + R'}{N - R'} > \frac{R + R'}{R - R'} \tag{153}$$

where

$$N = Q/4K\varepsilon \tag{154}$$

Now it can be easily shown that $R>N$: First, we write explicity

$$4+4s^2+8s^4 > 4/K + [4s^2(2\varepsilon+2K\varepsilon)]/4K\varepsilon + 8s^4 \tag{155}$$

Neglecting the small terms proportional to s^2 and s^4, we have:

$$4 > 4/K \tag{156}$$

This inequality holds for $K>1$, it becomes an equality for $K=1$, and reverses for $K<1$. Assuming that $K>1$, it is now evident that inequality (153) and, hence, (151) hold.

The question arises: What is the "correct" value of K? In EH theory, K is routinely taken to be within the range 1.5-2.0. This represents a reasonable range which "state of the art" computational calibrations are unlikely to challenge. It is then reasonable to assume that an analysis of stereoselection based on consideration of unnormalized H_{ij}'s is valid.

We can now summarize the important conclusions of the HL treatment of aromaticity as follows:

a) In the case of N-electron-M-center systems with $N = M \pm 1$, it is always possible to write a set of equivalent CW's differing in occupany by one spin orbital. These CW's are connected via interaction matrix elements which, to a first approximation, are proportional to the first power of AO overlap. The signs of the interaction matrix elements dictate whether the CW's form a Huckel or Möbius CW (<u>not AO</u>) array and this, in turn, determines the final energies of the corresponding eigenstates. We can say that aromaticity and antiaromaticity in such annulenes is primarily a function of the first power of AO overlap.

In the case of N-electron-N-center systems, it is always possible to write a set of equivalent CW's which now have identical spin orbital occupancy. These CW's are connected via interaction matrix elements which, to a first approximation, are proportional to the zero power of AO overlap. <u>Thus, to a first approximation, no distinction between aromatic and antiaromatic systems can be made on this basis.</u> However, a more detailed examination of the interaction matrix elements reveals that these are odd functions of AO overlap and that a distinction between aromatic and antiaromatic systems is possible on the basis of the smaller terms of the expansions. We can say that aromaticity and antiaromaticity in such annulenes is a function of high order AO overlap terms. Hence, stereoselection is predicted to operate strongly in N-electron-M-center and weakly in N-electron-N-center systems.

b) Stereoselection is a consequence of the mathematical form not only of the off diagonal but also of the diagonal matrix elements. With the exception of the two electron systems, the diagonal matrix elements are odd functions of AO overlap and, thus, they also play a role in determining whether a given system will be aromatic or antiaromatic.

c) A three orbital-two electron "bond" generated by cyclic in phase AO overlap is expected to be extremely strong on two counts: The CW's are devoid of any exchange repulsion and they also interact in a manner which defines a Hückel CW array. Only the second aspect is characteristic of an optimal three orbital-four electron "bond" and neither is characteristic of an optimal four orbital-four electron "bond".

The analysis of chemical stereoselection presented in this section is tantamount to a definition of the shortcomings of "resonance theory" which constitutes the quantum chemical foundation of the education of most chemists. According to this oversimplified version of VB theory, pi cyclopropenyl cation ($C_3H_3^+$) appears to be equally stable as pi cyclopropenyl anion ($C_3H_3^-$) since three equivalent low energy resonance structures can be drawn in each case. Similarly, pi benzene appears to be equally stable as pi cyclobutadiene because now two equivalent low energy resonance structures can be written in each case. These breakdowns occur because, in formulating "resonance theory", the signs and the magnitudes of the interaction matrix elements connecting the resonance structures were not considered in an explicit manner. Thus, $C_3H_3^+$ _appears_ as stable as $C_3H_3^-$ simply because we neglected the _signs_ of the interaction matrix elements and C_6H_6 appears as stable as C_4H_4 simply because we did not consider the _sizes_ of the interaction matrix elements.

Actually, the fact that "resonance theory" fails to differentiate between N-electron-N-center (Ne-Nc) aromatic and antiaromatic structures is due to an intriguing mishap which can be easily understood on the basis of this work. Specifically, we have already seen that one may define four different brands of qualitative HL theory, namely, AHL*, AHL°, AHL‡ and AHL theory, by adopting different integral approximations. Of these four brands, AHL* and AHL‡ theory are unique to the extent that they are the only approximate AHL methods which neglect high order AO overlap terms. Once these terms are neglected, no distinction between a Hückel and a Möbius Ne-Nc system can be made. Now, during the formative years of VB theory, the primary concern of the pioneers has been the illustration of the method. With this goal in mind, they opted for an approximate formulation of HL theory which would render the method comprehensible as well as applicable to model systems of interest. As a result, the brand of VB theory which became publicized in the chemical literature turned out to be HL theory in which high order AO overlap terms are neglected. We can denote this type of approximate theory by HL* and observe that it is different from all four approximate brands of HL theory formulated in this work. Furthermore, in contrast to AHL° and AHL theory, it fails to distinguish between Nc-Ne aromatic and antiaromatic systems and, indeed, it says that pi cyclobutadiene and pi benzene are equally stabilized by resonance. It is ironic that "resonance theory" was founded in part on HL*, rather than AHL° or AHL theory with the consequence that it ultimately failed to satisfy the curiosity of the experimentalist seeking a rationalization of the instability of cyclobutadiene and the remarkable stability of benzene and it paved the way towards the acceptance of HMO theory as the preeminent qualitative theoretical tool of chemistry despite its own significant failings. As we shall see in the following papers, many controversies would have been averted and our present state of understanding of molecular structure and reactivity would be much better had a different brand of approximate HL theory been selected as the illustrator of the key concepts of HL and VB theory.

A final cautionary remark: Chemical stereoselection can be predicted by HL theory as long as high order AO overlap terms are retained and regardless of the empirical approximations of integrals one may choose simply because the general matrix elements of two distinct stereochemical modes have the form:

$H = L + \delta E$ ("antiaromatic" geometry")

and

$H = L - \delta E$ ("aromatic" geometry)

By contrast, the energy difference between isolated reactants, e.g., 3 H_2, and corresponding cyclic complexes, e.g., H_6 hexagon, is dependent on the approximations of the theory because the general matrix elements now take the form indicated below.

$H = L \mp \delta L$ (reactants)

and

$H = L \mp \delta E$ (cyclic complex)

Since different approximations have different effect on δL and δE, the sign of the energy difference between the two molecular species depends on the brand of approximate theory employed although trends remain relatively unaffected.

I. The "Perfect" Form of Heitler-London Theory

One of the many reasons behind the unwillingness of chemists to explore VB theory as a qualitative theoretical tool is the apparent size of the CW basis set. For example, in the case of a 4e-4c system, a complete VB basis set includes just two HL CW's but eighteen more CW's of the "ionic" type. However, this pro1em can be circumvented by replacing localized by delocalized AO's and implementing HL theory over such a basis. This brand of HL theory is the implicit form of VB theory with a complete basis of CW's constructed from localized AO's. The only difference between this type of HL theory and the traditional VB theory is conceptual accessibility, i.e., explicit VB theory makes more chemical sense than implicit VB theory. In order to understand how this apparently unanticipated equivalence arises, it is necessary to consider the elementary problem of two electron-two center bonding, where the two orbitals are denoted by x_1 and x_2. The following discussion draws from important arly contributions by Coulson and Fischer[92] and Slater during the initial phase of development of VB theory.

The HL wavefunction of the ground state of a two electron-two orbital system of H_2, over a basis of x_1 and x_2 AO's is:

$$\Psi = |x_1\bar{x}_2| + |x_2\bar{x}_1| \tag{157}$$

We define a new basis of y_1 and y_2 AO's such that

$$y_1 = (x_1 + \lambda x_2)/(1 + 2\lambda s + \lambda^2)^{1/2} \tag{158}$$

$$y_2 = (x_2 + \lambda x_1)/(1 + 2\lambda s + \lambda^2)^{1/2} \tag{159}$$

$$s = \langle x_1|x_2\rangle \tag{160}$$

Henceforth, we denote all quantities over the basis of y_1 and y_2 by placing a bar on top of them. The HL wavefunction over this new basis is no more an ordinary HL wavefunction but rather a VB, or, $\overline{HL}$, wavefunction:

$$\bar{\Psi} = |y_1\bar{y}_2| + |y_2\bar{y}_1| \tag{161}$$

$$= (1+\lambda^2)[\underbrace{|x_1\bar{x}_2| + |x_2\bar{x}_1|}_{H\cdot \quad \cdot H}] + 2\lambda[\underbrace{|x_1\bar{x}_1|}_{H\overset{-}{:}\ H^+} + \underbrace{|x_2\bar{x}_2|}_{H^+:H^-}] \tag{162}$$

The overlap integral between y_1 and y_2 is

$$\bar{s} = \langle y_1 | y_2 \rangle = \frac{s+2\lambda/(1+\lambda^2)}{1-[2\lambda/(1+\lambda^2)]s} \tag{163}$$

We can distinguish the following important cases:

a) $\lambda = [-1 + (1 - s^2)^{1/2}]/s.$ (164)

In such a case, we obtain the antidelocalized VB wavefunction shown below:

$$\bar{\Psi} = (1+\lambda^2)(H\cdot\ \cdot H) - 2|\lambda|(H:^-H^+ + H^+H:^-) \tag{165}$$

For this value of λ, the AO's y_1 and y_2 are orthogonal and the $\bar{\Psi}$ describes an antibond.

b) $\lambda=0$. In such a case, we obtain the original HL wavefunction of (157).

c) $\lambda=1$. In such a case, we obtain overdelocalized VB wavefunction which is equivalent to the monodeterminantal MO wavefunction. Clearly, the "perfect" wavefunction is the one with $1>\lambda>0$.

Optimal two electron-two orbital bonding occurs when $\lambda = \bar{\lambda}$. Now, equations (162) and (163) both contain λ. Hence, bonding concepts can be extracted from any one of these two equations. Accordingly, we distinguish three equivalent descriptions of bonding due to delocalization:

a) The "localized AO description". This is nothing other than the traditional VB description of bonding. In this case, optimal bonding is gauged by the percent contribution of "covalent" and "ionic" VB CW's over localized AO's.

b) The "delocalized AO description" according to equation (161). This is the description used by Goddard in his formulation of the Orbital Phase Continuity Principle.[93] In this case, optimal bonding is gauged by the form of the delocalized AO's.

c) The "overlap description" according to equation (163). In this case optimal bonding is gauged by the overlap integral over delocalized AO's.

The effect of delocalization, i.e., the effect of replacing x_1 and x_2 by y_1 and y_2 in equation (157), or, the effect of making a transition from an HL to an $\overline{HL}$ description, on the relative energy of two systems which differ insofar as overlap is concerned can be easily predicted without the need of any computation if we reformulate appropriately the HL equations. Specifically, we note that local excitation and interatomic interaction can no longer be defined if we replace the HL by the $\overline{HL}$ AO's. This is a luxury afforded by HL theory in which we "know" the location of each electron by virtue of the fact that each AO is localized. Thus, we must reformulate the HL equations for the energy matrix elements in a suitable manner which allows a simple physical interpretation of delocalization. This can be easily achieved by reverting to equations (62, 67) and (71) and simply separating out monoelectronic, denoted now by the superscript 1, and bielectronic, denoted now by the superscript 2, components of F, G, and X as follows:

$$(F^1+G^1) = Y' \quad (166)$$

$$(F^2+G^2) = Y'' \quad (167)$$

$$X^1 \simeq X' \quad (168)$$

$$X^2 \simeq X'' \quad (169)$$

With these definitions, we obtain

$$H_{ii} \alpha \; Y'_i + Y''_i + X^1_i + X^2_i + \text{Nuclear Terms} \quad (170)$$

$$H_{ij} \alpha \; Y'_i + Y''_i + X^1_{ij} + X^2_{ij} + \text{Nuclear Terms} \quad (171)$$

$$H_{ij} \alpha X^1_{ij} + X^2_{ij} + \text{Nuclear Terms } (\Phi_i \text{ and } \Phi_j \text{ have no common } X_a\text{'s}) \quad (172)$$

The effect of delocalization on the energies of the H_{ij}'s can now be stated succintly as follows: As the localized HL are replaced by delocalized $\overline{HL}$ AO's, the sum of the primed monoelectronic terms becomes increasingly negative while the sum of the double primed bielectronic terms becomes increasingly positive. Initially, the rate of change of the monoelectronic energy component is greater than the rate of change of the bielectronic one. Ultimately, two electron destabilization catches up with one electron stabilization at which point optimal delocalization is achieved.

The above analysis can be best understood by reference to a specific example. Thus, in the space below, we show the HL energy expression for a two electron bond and we indicate by arrows some of the new critical terms which are introduced by replacing x_1 and x_2 by y_1 and y_2, respectively. Introduction of $\bar{\varepsilon}$ and replacement of βs by the much larger $\bar{\beta}\bar{s}$ term makes the energy more negative. By contrast, introduction of the self repulsion integral J_{11} has exactly the opposite effect. In graphic language, *we can say that two electron repulsion acts as the regulator of delocalization.* Henceforth, we shall be concerned with the monoelectronic delocalization recognizing the regulating action of the bielectronic terms.

$$E = \frac{1}{1+s^2}\Big(\underbrace{\overbrace{2\varepsilon + 2V}^{Y'}}_{\Downarrow\ \bar{\varepsilon}} + \underbrace{\overbrace{J_{12}}^{Y''}}_{\Downarrow\ J_{11}} + \underbrace{\overbrace{2\beta s}^{X^1}}_{\Downarrow\ \bar{\beta}\bar{s}} + \underbrace{\overbrace{K_{12}}^{X^2}}_{\Downarrow\ J_{11}}\Big) + V_{nn} \tag{173}$$

How does delocalization affect stereoselection? That is to say, what is the effect of delocalization on the aromatic-antiaromatic energy gap set at the HL theoretical level? In order to be able to answer this question, we must be able to ascertain the effect of AO delocalization on all four Y', Y", X^1, and X^2 terms, something which is by no means a simple task. Even if we were to neglect the effect of AO delocalization on the bielectronic terms of HL theory by assuming that the energy difference between an aromatic and an antiaromatic system reflects an overlap energy difference, the answer to the above question would still be difficult to obtain. The only viable approach is to compute explicitly the optimal delocalized AO's of the aromatic and the antiaromatic systems and, in the process, determine how their energy difference changes as a transition is made from HL to $\overline{\text{HL}}$ theory. With the optimal $\overline{\text{HL}}$ orbitals at hand, one can then develop a qualitative interpretation of the dependence of the aromatic-antiaromatic energy gap on the degree of delocalization permissible in the case of the specific system under consideration. Indeed, such an analysis of chemical stereoselection has been very elegantly carried out by

Goddard who pioneered Generalized VB (GVB) theory,[73g] the quantitative self-consistent version of $\overline{HL}$ theory, and applied it to a wide variety of chemically interesting problems. Using GVB theory, Goddard and his collaborators computed the optimal delocalized AO's of various prototypical systems in which localized AO's overlap in a cyclic manner and, on the basis of these calculations, enunciated the Orbital Phase Continuity Principle.[93]

$\overline{HL}$ theory contains the same information as "traditional" VB theory. However, the latter is much more chemically meaningful for reasons which are related to the scientific upbringing of chemists rather than to science itself. In particular, the schooling of most chemists is such that it renders them receptive to theoretical arguments involving, e.g., "covalent" and "ionic" "resonance structure" simply because they are thoroughly familiar with "resonance theory" and its applications to chemistry. Recognizing that "resonance theory" is the oversimplified version of VB theory, we can say that VB theory is more suitable to the background of most chemists than $\overline{HL}$ theory. In addition, $\overline{HL}$ theory presents an intellectual obstacle which, in an era of MO theory dominance, prohibits a widespread adoption. Specifically, the fact that the delocalized AO's of GVB theory are MO's generated by a VB-type theory creates conceptual difficulties for experimentalists who are accustomed to "pure", so to speak, MO or VB theory. It follows that in pursuing qualitative VB theory, we must either resign outselves to qualitative HL theory,or, we must go on further to adapt conventional VB theory in a way which makes it a useful tool for the analysis of the electronic structure of, at least, model systems.

J. Why HL Theory Cannot Qualify as a General Theory of Chemical Bonding

In a previous section, we argued that stereoselection, as exemplified by the relative energetics of a Hückel versus a Möbius system, can be treated in a "qualitative" sense via HL theory. The hallmark of problems of this type is that both systems under comparison can be described by one and the same set of HL CW's which are identically "connected", the expressions of the energy and overlap matrix elements have identical form in the two systems under comparison in a one to one sense, and the final energy difference is due to a difference in the signs of the AO overlap integrals. As a result, the final answer does not depend on the precise numerical value attained by the critical variable, namely, the AO overlap integral, but, rather, it depends on whether this critical variable is positive, or negative. To put it crudely, in problems of this type, it is parity, i.e., the sign rather than magnitude of s_{ij}, which holds the key to the understanding of the problem. As a result, HL theory is "qualitatively" adequate.

The situation becomes radically different when we deal with practically every other problem of chemistry. For example, the two systems of four AO's and two electrons shown below cannot be described by one and the same set of HL CW's which are identically "connected". This is made evident in Figure 6 which shows that the interaction pattern of the six CW's is different in the two systems. Note that, as the two systems do not involve a cyclic AO overlap, the HL energy expressions should be independent of the signs of AO overlap integrals. That this is the case in made evident in Figure 6 which shows that, though cyclic CW overlap obtains in both systems, the sign of the product of the CW overlap integrals is independent of the signs of the s_{tu}'s since it is a function of s_{tu}^2.

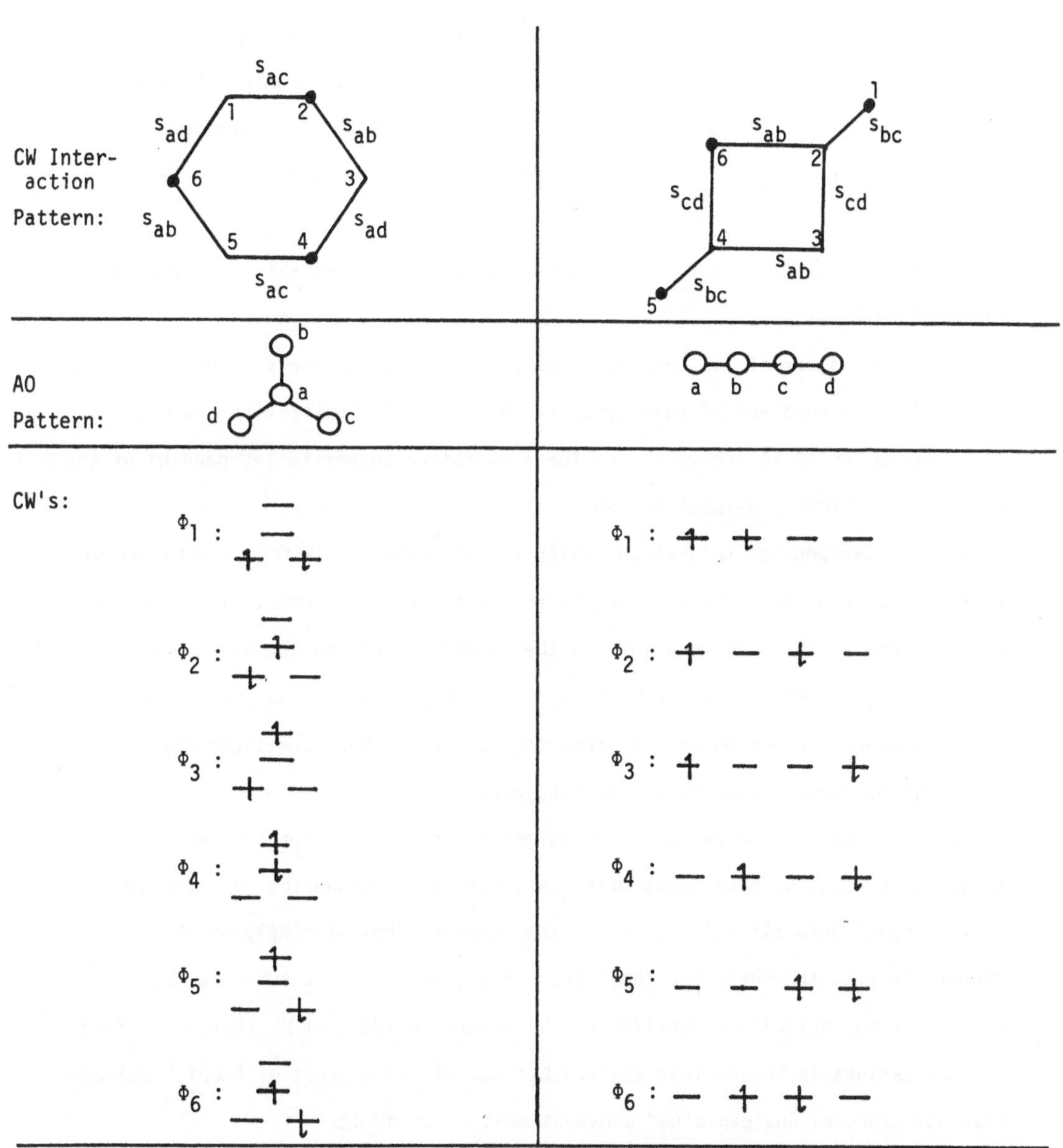

FIGURE 6: CW interaction in linear and Y-shaped 4 electron-4 orbital systems. Φ_i's represent CW's, s_{tu}'s AO overlap integrals, and S_{ij} is proportional to s_{tu} in the manner indicated on the sides of the polygons. Black circles indicate the HL CW's.

We are now faced with a problem which is completely different from the previous one. For, since the HL CW's are not identically "connected", the energy and overlap matrices do not have identical form and the final energy difference is no longer due to a difference in the signs of the AO overlap integrals.

On the basis of the above discussion, we can define two extremely different chemical problems:

a) Comparisons of two systems which are described by the same network of HL CW's. In many problems of this type, HL theory is the qualitative analogue of the rigorous $\overline{\text{HL}}$ or VB theories, and the qualitative answer is independent of the magnitude of "ionic" delocalization.

b) Comparisons of two systems which are described by different networks of HL CW's. In problems of this type, HL theory is never the qualitative analogue of the rigorous $\overline{\text{HL}}$ or VB theories and the answer depends on the magnitude of "ionic" delocalization. In problems of this type,either of two possible answers can be obtained depending on the electronic nature of the constitutent atoms. This will be demonstrated in a following paper.

A second problem arises when the target systems are no longer homonuclear systems, as those we have dealt with, but, rather, heteronuclear systems in which "ionic" delocalization , i.e., the contributions of singly, doubly, etc., "ionic" CW's to the total wavefunctions, is significant. In such a case, HL theory becomes definitely unreliable. We hasten to add that HL theory <u>can</u> fail in applications to homonuclear systems because of the neglect of "ionic" delocalization for reasons explained above as well as on p. 95.

On the basis of the above considerations, it is evident that we are justified to pursue a VB rather than HL theory of chemical bonding as a result of the limitations of HL theory. Furthermore, we are justified to pursue a VB rather than an equivalent $\overline{HL}$ theory of chemical bonding for two main reasons:

a) It offers conceptual advantages whenever we deal with problems where we want to find out how environmental perturbations, i.e., substituent effects, solvent effects, etc., act as to modify the electronic structure of the unperturbed system.

b) It projects electron delocalization in a way which can be exploited in order to develop a qualitative VB theory of chemical bonding of polyelectronic systems, such as the one presented in the second part of this work.

K. The Transition to "Orthogonal" VB Theory and Its Approximate Variants

In the HL and $\overline{HL}$ theories discussed above, all expressions are functions of AO overlap. The reader who is not very familiar with VB formalism may be inclined to think that neglect of overlap in any shape or form would be catastrophic. That this is not the case and that VB theory can be developed starting with an orthogonal AO basis are well known facts to theoretical chemists[94] but not to practicing experimentalists. Hence, in the space below, we provide a discussion of these matters for the purpose of maintaining continuity as well as generating a comprehensive overview of VB and MO theories and their interrelationships. On this basis, one can finally understand the fallacies which have come to being from inappropriate utilization of computational methods and incorrect interpretation of computational results. We shall illustrate the salient points by reference to the simple two electron-two orbital diatomic H-H.

Starting with a basis set of conventional nonorthogonal AO's, the HL ground wavefunction of H_2 is:

$$\Psi_{HL} \simeq H\cdot \;\; \cdot H \tag{174}$$

We say that, at this level of theory, the ground wavefunction is <u>localized</u>. The "perfect" VB ground wavefunction is generated by addition of the appropriate "ionic" CW's which describe delocalization and it has the following form:

$$\Psi_{VB} = H\cdot \;\; \cdot H + \lambda(H{:}^{-}H^{+} + H^{+}H{:}^{-}) \qquad \lambda<1 \tag{175}$$

We say that, at this level of theory, the ground wavefunction is <u>optimally delocalized.</u>

Alternatively, the same result can be obtained at the level of $\overline{HL}$ theory where now the basis AO's are not localized nonorthogonal but rather delocalized nonorthogonal AO's. Finally, the monodeterminantal MO ground wavefunction has the form:

$$\Psi_{MO} = H\cdot \;\cdot H + \lambda(\bar{H}\!: H^+ + H^+ :H^-) \qquad \lambda=1 \tag{176}$$

We say that the MO ground wavefunction is <u>hyperdelocalized</u>, i.e., the contribution of the "ionic" CW's is exaggerated to an extent which becomes energetically counterproductive.

Let us now examine what happens if we switch to an orthogonal AO basis. The HL ground wavefunction no longer defines a "covalent" two-electron bond "joining" the two H atoms but rather a two-electron <u>antibond</u>. In this case, the basis AO's are not delocalized nonorthogonal AO's, as in $\overline{HL}$, theory but rather antidelocalized nonorthogonal AO's as is evident from the explicit form of this type of ground wavefunction:

$$\tilde{\Psi}_{HL} = H\cdot \;\cdot H - \lambda(H\!:^-H^+ + H^+H\!:^-) \tag{177}$$

We can symbolize this brand of theory by $\widetilde{HL}$. The perfect $\widetilde{VB}$ ground wavefunction is now generated by addition of the appropriate "ionic" CW's but with a much larger coefficient than in the case of VB theory in order to counteract the antidelocalization generated at the level of $\widetilde{HL}$ theory. Accordingly, VB, $\overline{HL}$, and $\widetilde{VB}$ theories are all equivalent "perfect" theories, with HL theory being an excellent approximation in the case of systems held together by nonpolar bonds. However, $\widetilde{HL}$ theory is a nontheory. In formal terminology, we say that $\widetilde{HL}$ is a nontheory because the HL ground wavefunction is not invariant to an orthogonal transformation of the AO basis and such a transformation has the consequences made evident by equation (177).

The above discussion makes clear that if a complete basis of VB CW's is used, departure from orthogonal or nonorthogonal AO basis is immaterial insofar as VB theory (not HL theory) is concerned. Hence, a set of approximate versions of VB theory can be rigorously formulated on the basis of the integral approximations discussed in a previous section.

L. The Physical Interpretation of VB Configuration Wavefunctions

One of the main reasons why VB theory has never been developed into a powerful qualitative theory of chemical bonding lies in the fact that the CW basis has been regarded as a short, long, or interminable, depending on the system at hand, list of functions which cannot be classified in a physically meaningful manner which projects their interrelationships and the role that they play in determining the relative energies of ground and excited states of one or more systems. In this section we develop a physical model for the interpretation of VB wavefunctions which is based on three novel constructs:

a) The concept of the elementary bond.

b) The concept of the elementary structure.

c) The concept of intrinsic, direct, and indirect coupling between elementary bonds and structures.

We shall describe this model by departing from the traditional concepts of HL theory and developing the key ideas in modest detail. Since concepts are best understood by reference to a specific example, we have chosen the four electron-three orbital pi system of the allyl anion as the illustrator of our approach.

The HL theoretical description of a two-electron two-orbital bond is a single HL CW. Addition of two "ionic" CW's produces the corresponding VB theoretical description. Thus, for example, the HL and VB theoretical descriptions of the electron pair bond of H_2 are as indicated below, where x_1 and x_2 stand for the two 1s hydrogen AO's and the symbols in parenthesis denote how the electrons are allocated to different AO's in each CW.

HL: $H\cdot\ \cdot H\ (x_1x_2)$

VB: $H^+\ H\colon^-\ (x_2^2) \leftrightarrow H\cdot\ \cdot H(x_1x_2) \leftrightarrow H\colon^-\ H^+\ (x_1^2)$

Henceforth, we define a two electron-two orbital bond described by the optimal set of three VB CW's such as the ones shown above as an elementary bond.

The HL theoretical description of pi allyl anion is shown below.

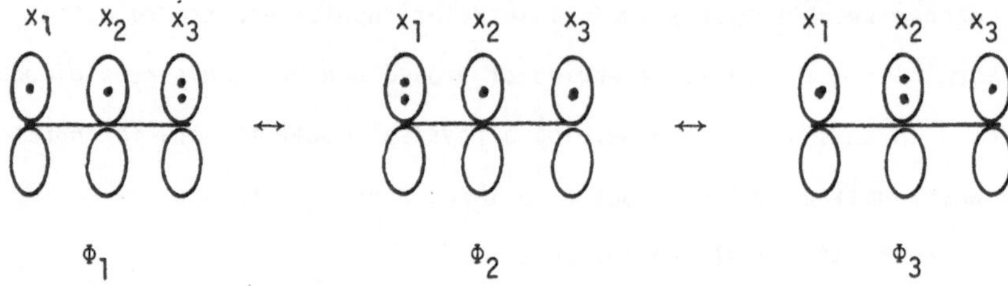

We can generate the corresponding VB description by replacing each HL electron pair bond by the corresponding VB elementary bond. In doing so, each of the three HL CW's shown above becomes the progenitor of three VB CW's which can be represented in matrix form as indicated below, where x_1, x_2, and x_3 represent the three pi type AO's of the allyl anion.

Φ_1 Generator

	x_1^2	x_1x_2	x_2^2
x_3^2	Φ_4	Φ_1	Φ_5

Φ_2 Generator

	x_2^2	x_2x_3	x_3^2
x_1^2	Φ_6	Φ_2	Φ_4

Φ_3 Generator

	x_1^2	x_1x_3	x_3^2
x_2^2	Φ_6	Φ_3	Φ_5

A pictorial representation of each of the three sets of VB CW's generated by the three HL CW's is given below.

	Φ_1 Generator		Φ_2 Generator		Φ_3 Generator
Φ_4	++ — ++	Φ_6	++ ++ —		++ ++ —
Φ_1	+ + ++	Φ_2	++ + +	Φ_3	+ ++ +
Φ_5	— ++ ++	Φ_4	++ — ++		— ++ ++

Recall now that, in traditional HL theory, each HL CW is referred to as a "bond eigenfunction," or, a "resonance structure." We now convene that the optimal set of VB CW's generated from an HL CW by substituting an elementary HL by an elementary VB bond be termed an <u>elementary structure</u>. Clearly, we will always have as many elementary structures as HL CW's. The pictorial and formal representations of the three possible elementary structures of the pi allyl anion system are given below:

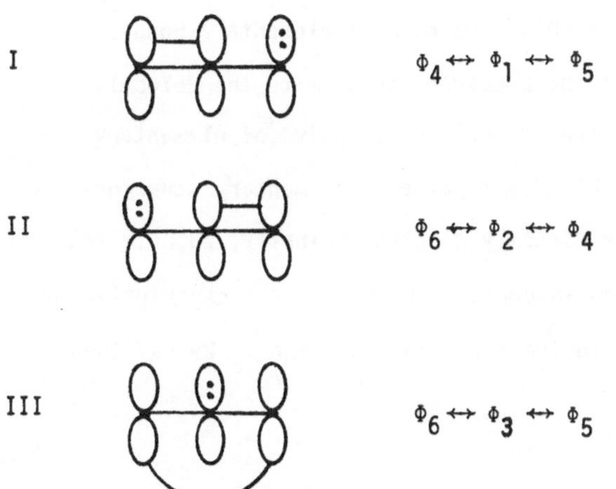

The procedure outlined above for the generation of the three elementary structures is also a procedure for generating the CW basis required for the treatment of the pi allyl anion. We note that in proceeding in the way we did, we have generated nine CW's with three of them being redundant. The remaining six constitute a complete set of CW's. In this case, we say that the CW basis is spanned by the elementary structures. This situation is rather rare. In most cases, the complete CW basis is not spanned by the elementary structures, i.e., there exist CW's which cannot be generated by the "basis forming machine" specified above.

One of the important features of the analysis presented above is that it leads to a qualitative understanding of how elementary bonds and structures interact with each other through delocalization. In the absence of symmetry constraints, we can distinguish three types of bond coupling: <u>intrinsic, direct, and indirect coupling.</u> Let us see exactly how such couplings arise. Consider the bond structures I and II of the pi allyl anion system. We ask the question: How are these two elementary structures coupled? First, we note that they do share a common VB CW, namely, Φ_4. We say then that I and II are intrinsically coupled. The physical significance of intrinsic coupling is self evident. In delocalizing over a set of AO's, two or more elementary bonds interact because the delocalization of one precludes or permits the delocalization of the other. Second, we note that VB CW's descriptive of elementary structure I can interact with the VB CW's descriptive of elementary structure II. We say then that the I and II are directly coupled. Finally, we note that the VB CW's descriptive of I and II can interact with the VB CW's descriptive of III and, thus, they can also interact indirectly with each other. We say then that I and II are indirectly coupled.

With the above in mind, we can now group VB CW's in a way which reflects the existence of elementary bonds and structures. Thus, we can subdivide the VB CW's into the following classes.

a) Unique CW's, i.e., CW's which are unique to a given elementary structure.

b) Common CW's, i.e., CW's which are common to two or more elementary structures.

c) Extrinsic CW's, i.e., CW's other than those generated by the method described above and needed for the completion of the CW basis set.

We shall see that with these definitions in mind, we shall be able to develop a clear idea as to how stereoselection arises in various cyclic systems.

M. Hückel Valence Bond Theory of Stereoselection

Why should we pursue an understanding of aromaticity and antiaromaticity on the basis of HVB theory? The first and most important reason is that HVB theory is equivalent to the theory which is the foundation of most present day concepts of chemistry: HMO theory with neglect of overlap. Secondly, an HVB treatment of steroselection has never been presented before. Accordingly, this section fills a void in the theoretical literature. Thirdly, the HVB theory of model systems constitutes preparation for the VB theory of real systems to be presented in following papers. Finally, HVB theory is the simplest illustrator of the conceptual advantages of the VB method, in a general sense. It is the first obligatory step along the pedagogical route leading to ultimate mastery of rigorous VB theory.

We begin with a comparison of the HVB and HMO methods as applied to a two electron-two orbital system, such as the pi bond of ethylene. In HMO theory, one starts with a trial wavefunction which is a linear combination of two pi AO's and solves a 2x2 secular determinant to obtain two MO's which can now be used to generate the states of the pi system of ethylene. In HVB theory, one starts with a trial wavefunction which is a linear combination of three VB CW's and solves a 3x3 secular determinant in order to obtain the states directly. We note the following:

a) The HMO method uses a smaller number of basis functions than the HVB method. Actually, this difference grows sharply as the number of AO's and electrons increases. Thus, there is an "economy" factor operating in favor of the HMO method.

b) The HVB method produces an explicit representation of the electron distribution within each state. In turn, this leads us to anticipate directly the effect of interelectronic repulsion on the energy of each state. There is no such explicit warning mechanism in HMO theory. In other words, in using the integral approximations of HMO theory, we commit the error of neglecting interelectronic repulsion. In HVB theory we are starkly aware of this error and its implications while in HMO theory we have lost track of it.

Next, we outline the formal HVB procedure to be used in the rest of this work.

HVB theory constitutes the simplest introduction to rigorous VB theory. This convenience is afforded by the fact that the energy and overlap matrix elements over VB CW's assume an exceedingly simple form, as indicated below:

$$H_{ii} = \Sigma\, \alpha_t \tag{178}$$

$$H_{ij} = K\beta_{tu} \tag{179}$$

$$S_{ij} = 0 \tag{180}$$

In the above expressions, the indices i and j refer to VB CW's, t and u to AO's, K is a factor, and α and β have the same meaning as in HMO theory. The summation in equation (178) is over all occupied spin orbitals. With this background and the concepts developed in the previous section, we can perform an HVB analysis of a problem in a manner which sheds light on the innermost secrets of chemical bonding. In our study of stereoselection in cyclic systems, we shall use the following standard procedure:

a) The elementary structures necessary for the treatment of the problem at hand are defined.

b) The VB CW basis set is partitioned into unique, common, and extrinsic CW's.

c) The electronic basis of stereoselection is investigated by performing sequential computations wherein the basis set is augumented in a way which can reveal what types of CW's are responsible for stereoselection.

4e-3c System

The HL treatment of a 4e-3c system requires three HL CW's. The VB treatment of the same system requires three elementary structures generated from the three HL CW's. The six linearly independent CW's which are generated in this manner are shown in Figure 7a. They constitute a complete CW basis set which can be partitioned in the manner indicated in Figure 7b. The complete energy matrix is shown in diagrammatic form in Figure 8 and the results of the stepwise construction of the eigenstates as prescribed before are shown in Figure 9.

First, we note two interesting aspects of the CW basis set:

a) All unique CW's are of the HL variety and all common CW's are of the "ionic" variety.

b) The interaction of HL CW's defines "covalent" delocalization while the interaction of the HL with the common "ionic" CW's defines "ionic" delocalization.

Secondly, we note two interesting aspects of the energy matrix:

a) All six CW's have the same energy at the level of HVB theory. This is, of course, incorrect since the "ionic" CW's "suffer" much more from interelectronic repulsion than the HL CW's. We make note of this fact for future consideration. Typically, we have:

$$\langle\Phi_1|\hat{H}|\Phi_1\rangle = \alpha_1 + \alpha_2 + 2\alpha_3 = 4\alpha \tag{181}$$

$$\alpha_1 = \alpha_2 = \alpha_3 = \alpha_4 \tag{182}$$

b) It is clear that each elementary bond is coupled to the other intrinsically, directly, and indirectly. For example, the elementary structures I and II (See Figure 7a) share the common CW Φ_4, are connected directly via the H_{12} matrix element and indirectly via the H_{13} and H_{23} matrix elements.

(a)

Elementary Structure	Pictorial Representation	Constituent CW's		
I	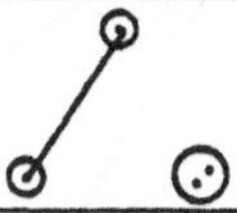	Φ_1	Φ_4	Φ_5
II		Φ_2	Φ_4	Φ_6
III	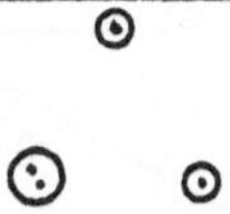	Φ_3	Φ_5	Φ_6

(b)

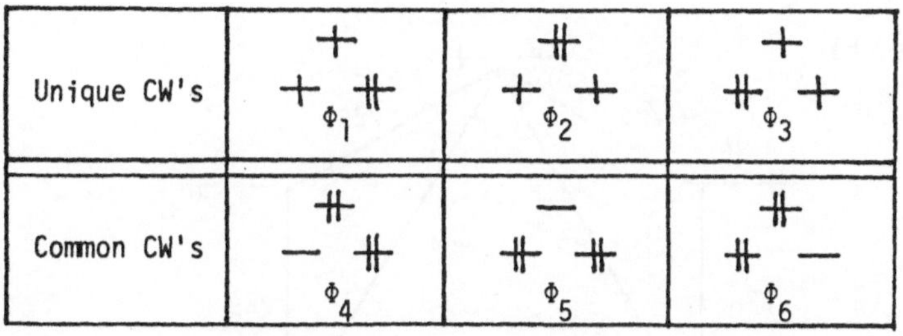

Unique CW's	Φ_1	Φ_2	Φ_3
Common CW's	Φ_4	Φ_5	Φ_6

Figure 7: a) Elementary structures and their constituent CW's for the 4e-3c system.

b) Classification of CW's of the 4e-3c system.

a)

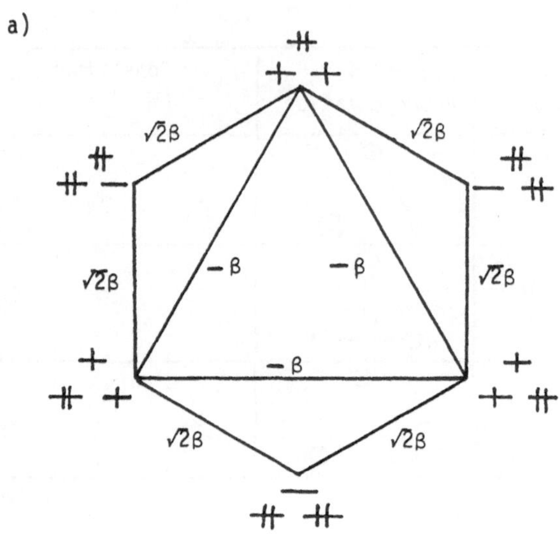

b)

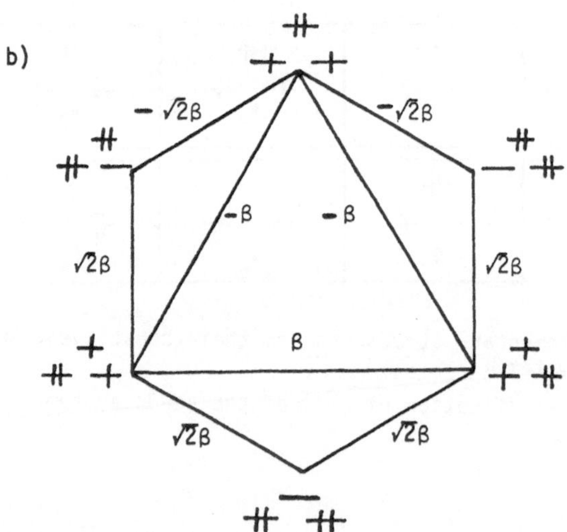

Figure 8: Pictorial representation of the HVB interaction energy matrix for the 4e-3c system. (a) Hückel AO system. (b) Möbius AO system.

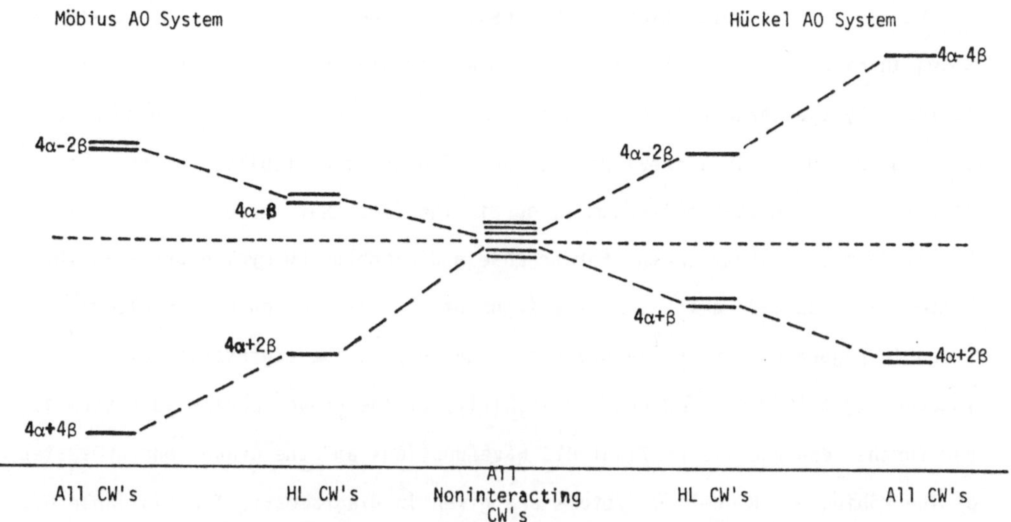

Figure 9: Stepwise construction of the three lowest energy Hückel and Möbius HVB eigenstates having principally HL character for the 4e - 3c system.

We now turn our attention to the results of the HVB computations which are shown in Figure 9. In the first stage, we compute the eigenstates by interacting only the three unique HL CW's. This is an approximate HL treatment of the type discussed in a previous section. The greater stability of the Möbius relative to the Hückel AO system is due to the fact that three HL CW's define a Hückel CW cyclic array in the former case and a Möbius CW cyclic array in the latter case, as made evident by the signs of the interaction matrix elements shown in Figure 8. In the second stage, we compute the eigenstates by interacting all CW's. The greater stability of the ground Möbius AO system is now further enhanced. The final HVB wavefunctions and the ground energy states of the Möbius and Hückel AO systems are given in diagrammatic form in Table 6.

Some of the more noteworthy trends are the following:

a) The state manifolds of the Möbius and Hückel AO systems have an inverse relationship, as seen by inspection of Figure 9. This is simply due to the fact that the two systems differ with respect to the sign of an AO overlap integral. As a result, a bonding state of one system is transformed into a corresponding antibonding state of the other.

b) Because of the initial splitting of the HL states, three of the final six states end up being primarily "covalent" (66%) and the other three end up being primarily "ionic".

c) The three lowest lying degenerate states of the Hückel AO system are known in MO parlance as "diradical" states. It is immediately obvious that the term "diradical" is misleading since all three states have a mix of "radical" and "zwitterionic" character, the former due to the HL CW's and the latter to the "ionic" CW's. At the level of rigorous VB theory, we expect that the three-fold degeneracy will be lifted simply because one of the three states is

Table 6. HVB and NDO-VB Wavefunctions for Hückel and Möbius 4 electron - 3 center Systems.*

		Ψ_i	Energy	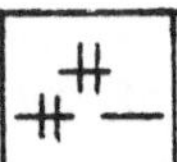 ‡ / ‡ —	 — / ‡ ‡	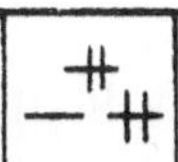 ‡ / — ‡	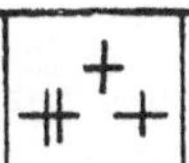 + / ‡ +	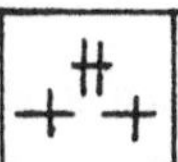 ‡ / + +	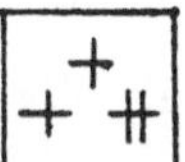 + / + ‡
HVB	Hückel AO System	Ψ_1	2.0	0	.42	-.40	.56	.59	0
		Ψ_2	2.0	-.47	.22	.25	-.35	.31	.66
		Ψ_3	2.0	.47	.47	.47	.33	-.33	.33
	Möbius AO System	Ψ_1	4.0	.33	.33	.33	-.47	-.47	.47
NDO-VB	Hückel AO System	Ψ_1	-7.5	.16	.16	-.32	.75	.37	-.37
		Ψ_2	-7.5	-.27	.27	0	0	.65	.65
		Ψ_3	-4.5	-.42	.42	.42	.40	.40	.40
	Möbius AO System	Ψ_1	-16.4	-.27	-.27	-.27	.51	.51	-.51

*NDO-VB energies in eV and HVB energies in β units assuming $\alpha = 0$. For NDO-VB computations, see Appendix 2.

more "ionic" than the other two and inclusion of electron-electron interaction will raise the energy of Ψ_3 relative to Ψ_1 and Ψ_2, the latter two being degenerate E states.

d) Figure 9 shows that in the case of the aromatic Möbius AO system, one half of the binding energy of the ground state (Ψ_1) is due to the interaction of the HL CW's between themselves and the other half to the interaction of the HL with the "ionic" CW's. The same is exactly true for the binding energy of the triply degenerate Hückel AO ground state. In other words, "ionic" delocalization lowers the energies of both the aromatic and antiaromatic systems exactly as anticipated on the basis of the $\overline{HL}$ analysis of chemical bonding. This energy lowering is exaggerated because of the neglect of coulomb and exchange correlation. At the level of rigorous VB theory, we expect that inclusion of electron-electron interaction will raise the energies of the "ionic" CW's relative to the energies of the HL CW's, thus diminishing the interaction of the two sets of CW's. As a result, "covalent" delocalization, <u>due to the interaction of the HL CW's</u>, will be rendered much more important than "ionic" delocalization, <u>due to the interaction of the HL and the "ionic" CW's</u>. Accordingly, most of the binding of the ground state of either aromatic or antiaromatic system will be due to "covalent" delocalization.

e) The lower energy of the ground state of the Möbius AO system compared to that of the Hückel AO system is due to the fact that "covalent" and "ionic" delocalizations are both twice as strong in the case of the aromatic system. Thus, Figure 9 shows that the interaction of the HL CW's leads to a Möbius AO ground state which has a binding energy twice as large as that of a Hückel AO ground state. Addition of the "ionic" CW's further enhances the greater stability of the former relative to the latter by a proportional amount. The effect of the addition of the "ionic' CW's to the basis set on the relative energies of the Hückel and Möbius AO systems is entirely predictable from examination of the signs of the CW interaction matrix elements in Figure 8. Specifically, addition of the three equivalent "ionic" CW's generates three additional cyclic Hückel CW arrays in the case of the Möbius AO system while it produces three additional cyclic Möbius CW arrays in the case of the Hückel AO system. Thus, the effect of "ionic" delocalization is merely to enhance the sterochemical preference of the 4e-3c system due to "covalent" delocalization.

4e-4c System

The HL treatment of a 4e-4c system requires two HL CW's. The VB treatment of the same problem requires two elementary structures generated from the two Kekule CW's which describe two-electron bonds in the manner indicated below:

In contrast to the previous case, the sixteen CW's generated in this manner do not constitute a complete CW basis set. This can only be achieved by addition of four CW's, each of which describes an electron transfer from one elementary structure to the other. Their characteristic feature is a doubly occupied AO sandwiched between two singly occupied ones. The partitioned CW basis set is shown in Figure 10 and the complete energy matrix is given in Table 7. It is evidence that the CW basis set required for the treatment of a 4e-4c system has totally different characteristics from that required for the treatment of a 4e-3c system. Thus, at the level of HVB theory, the HL CW's do not interact with each other and "covalent" delocalization is now defined by the interaction of elementary and extrinsic CW's.

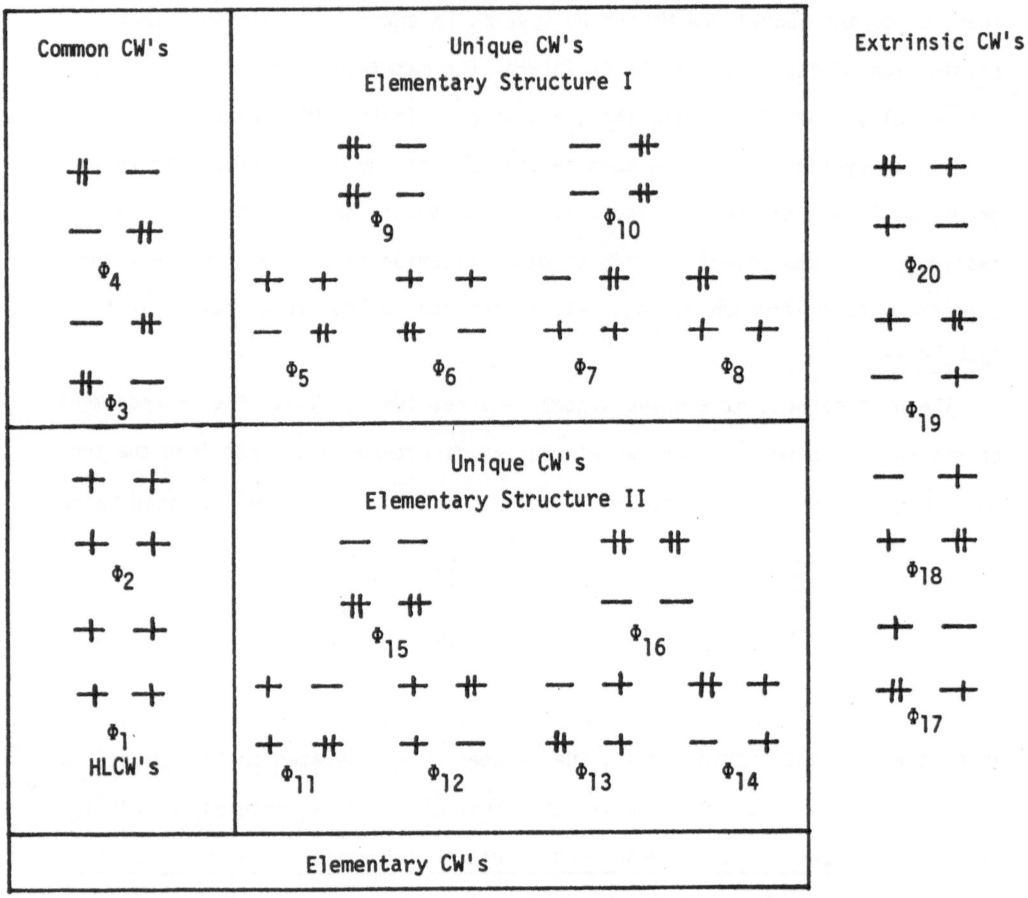

Figure 10: Classification of CW's for the 4c-4e system.

Table 7. Full Energy Matrix for 4c-4e$^-$ System

	+ + / + +	+ + / + +	# + / − +	# + / + −	# − / + +	+ # / + −	+ # / − +	− # / + +	− + / # +	+ − / # +	+ + / # −	+ − / + #	− + / + #	+ + / − #	# # / − −	− − / # #	# − / # −	− # / − #	# − / − #	− # / # −
+ + / + +	0																			
+ + / + +	0	0																		
# + / − +	-1.41	0	0																	
# + / + −	0	0	1.0	0																
# − / + +	.71	1.22	0	1.0	0															
+ # / + −	-1.41	0	0	-1.0	0	0														
+ # / − +	0	0	-1.0	0	0	1.0	0													
− # / + +	.71	1.22	0	0	0	0	1.0	0												
− + / # +	-1.41	0	0	0	0	0	0	0	0											
+ − / # +	0	0	0	0	-1.0	0	0	0	1.0	0										
+ + / # −	.71	1.22	0	-1.0	0	0	0	0	0	1.0	0									
+ − / + #	-1.41	0	0	0	0	0	0	0	0	-1.0	0	0								
− + / + #	0	0	0	0	0	0	0	-1.0	-1.0	0	0	1.0	0							
+ + / − #	.71	1.22	0	0	0	0	-1.0	0	0	0	0	0	1.0	0						
# # / − −	0	0	1.41	0	0	1.41	0	0	0	0	0	0	0	0	0					
− − / # #	0	0	0	0	0	0	0	0	1.41	0	0	1.41	0	0	0	0				
# − / # −	0	0	0	0	1.41	0	0	0	0	0	1.41	0	0	0	0	0	0			
− # / − #	0	0	0	0	0	0	0	1.41	0	0	0	0	0	1.41	0	0	0	0		
# − / − #	0	0	1.41	0	1.41	0	0	0	0	0	0	1.41	0	1.41	0	0	0	0	0	
− # / # −	0	0	0	0	0	1.41	0	1.41	1.41	0	1.41	0	0	0	0	0	0	0	0	0

Table 8. HVB and NDO-VB Wavefunctions for Hückel and Möbius 4 electron - 4 center Systems*

		Ψ_i	E_i	+ + + +	+ + + +	− + ‡ +	+ ‡ + −	+ + − ‡	+ + ‡ −	‡ − + +	‡ + − +	− ‡ + +	+ − + ‡	+ − ‡ +	‡ + + −	+ ‡ − +	− + + ‡	− − ‡ ‡	‡ ‡ − −	− ‡ − ‡	‡ − ‡ −	− ‡ ‡ −	‡ − − ‡
HVB	Hückel AO System	Ψ_1	4.0	.25	.43	0	0	.35	.35	.35	0	.35	0	0	0	0	0	0	0	.25	.25	.25	.25
		Ψ_2	4.0	.50	0	-.35	-.35	0	0	0	-.35	0	-.35	0	0	0	0	-.25	-.25	0	0	-.25	-.25
		Ψ_3	4.0	0	0	-.25	-.25	.25	-.25	.25	.25	-.25	.25	-.25	.25	-.25	.25	0	0	0	0	.35	.35
		Ψ^+	4.0	.53	.29	-.25	-.25	.25	.25	.25	-.25	.25	-25	0	0	0	0	-.17	-.17	.17	.17	0	0
		Ψ^-	4.0	.18	-.30	-.25	-.25	-.25	-.25	-.25	-.25	-.25	-25	0	0	0	0	-.17	-.17	-.17	-17	-.35	-35
	Möbius	Ψ_1	5.6	.37	.21	-.25	.25	.25	.25	.25	-.25	.25	.25	-.17	-.17	.17	.17	.13	.13	.13	.13	.25	.25
NDO-VB	Hückel	Ψ_1	-13.7	.65	.38	-.23	-.23	.23	.23	.23	-.23	.23	-.23	0	0	0	0	0	0	0	0	0	0
		Ψ_2	-10.8	-.29	.51	.24	.24	.24	.24	.24	.24	.24	.24	0	0	0	0	.10	.10	.10	.10	.28	.28
		Ψ_3	-9.3	0	0	-.26	-.26	.26	-.26	.26	.26	-.26	.26	-.25	.25	-.25	.25	0	0	0	0	-.33	.33
	Möbius	Ψ_1	-19.3	.50	.29	-.24	.24	.24	.24	.24	-.24	.24	.24	-.15	-.15	.15	.15	0	0	0	0	.21	.21

*NDO-VB energies in eV and HVB energies in β units assuming α = 0. For NDO-VB computations, see Appendix 2.

We now discuss various theoretically interesting points:

a) The three lowest lying states of the Hückel AO system are degenerate at the level of HVB theory. However, the explicit forms of the HVB wavefunctions allow us to anticipate precisely what is going to happen at the level of rigorous VB theory. Thus, if we replace Ψ_1 and Ψ_2 by their sum and difference, we obtain the wavefunctions Ψ^+ and Ψ^- shown in Table 8. Now, Ψ^+, Ψ^-, and Ψ_3 define the triply degenerate ground state. The percentage ionic character of these three HVB states is very different with Ψ^+ being the least "ionic" and Ψ_3 being the most "ionic" as it is totally devoid of any contribution from the HL CW's. Accordingly we expect that at the level of rigorous VB theory, the triple degeneracy will be lifted with the relative energies of Ψ^+, Ψ^-, and Ψ_3 becoming $E(\Psi_3)>E(\Psi^-)>E(\Psi^+)$. This is exactly what ab initio SCF-MO-CI calculations show.

b) The three lowest lying degenerate states of the Hückel AO system correspond to elementary structure I(Ψ_1), elementary structure II(Ψ_2), and some hybrid state made up of "ionic" elementary and extrinsic CW's (Ψ_3). The first two states are the "parents" of the diradical intermediates in nonpolar cycloadditions and the third is the "parent" of zwitterionic intermediates in highly polar cycloadditions. By contrast, the lowest state of the Möbius AO system corresponds to elementary structures I plus II coupled via the extrinsic CW's in a monoelectronic sense. The tabular representation of the ground states of the Möbius and Hückel AO systems (Table 8) is the most detailed, yet pictorial, description of aromaticity and antiaromaticity in even center annulenes.

c) It is not an exaggeration to say that the most eminently successful qualitative theoretical model is the FO-PMO model. Professor Dewar of Texas has been one of its early advocates and the number of chemists who find this

Table 8. HVB and NDO-VB Wavefunctions for Hückel and Möbius 4 electron - 4 center Systems*

		Ψ_i	E_i	+ + + +	+ + + +	− + ⧺ +	+ ⧺ + −	+ + − ⧺	+ + ⧺ −	⧺ − + +	⧺ + − +	− ⧺ + +	+ − + ⧺	+ − ⧺ +	⧺ + + −	+ ⧺ − +	− + + ⧺	− − ⧺ ⧺	⧺ ⧺ − −	− ⧺ − ⧺	⧺ − ⧺ −	− ⧺ ⧺ −	⧺ − − ⧺
HVB	Hückel AO System	Ψ_1	4.0	.25	.43	0	0	.35	.35	.35	0	.35	0	0	0	0	0	0	0	.25	.25	.25	.25
		Ψ_2	4.0	.50	0	-.35	-.35	0	0	0	-.35	0	-.35	0	0	0	0	-.25	-.25	0	0	-.25	-.25
		Ψ_3	4.0	0	0	-.25	-.25	.25	-.25	.25	.25	-.25	.25	-.25	.25	-.25	.25	0	0	0	0	.35	.35
		Ψ^+	4.0	.20	.29	-.23	-.23	.23	.23	.23	-.23	.23	-.23	0	0	0	0	-.17	-.17	.17	.17	0	0
		Ψ^-	4.0	-.10	-.22	-.18	-.18	-.18	-.18	-.18	-.18	-.18	-.18	0	0	0	0	-.13	-.13	-.13	-.13	-.26	-.26
	Möbius	Ψ_1	5.6	.37	.21	-.25	.25	.25	.25	.25	-.25	.25	.25	-.17	-.17	.17	.17	.13	.13	.13	.13	.25	.25
NDO-VB	Hückel	Ψ_1	-13.7	.65	.38	-.23	-.23	.23	.23	.23	-.23	.23	-.23	0	0	0	0	0	0	0	0	0	0
		Ψ_2	-10.8	-.29	.51	.24	.24	.24	.24	.24	.24	.24	.24	0	0	0	0	.10	.10	.10	.10	.28	.28
		Ψ_3	-9.3	0	0	-.26	-.26	.26	-.26	.26	.26	-.26	.26	-.25	.25	-.25	.25	0	0	0	0	-.33	.33
	Möbius	Ψ_1	-19.3	.50	.29	-.24	.24	.24	.24	.24	-.24	.24	.24	-.15	-.15	.15	.15	0	0	0	0	.21	.21

*NDO-VB energies in eV and HVB energies in β units assuming α = 0. For NDO-VB computations, see Appendix 2.

approach useful in a pedagogical as well as in an operational sense continues to grow. FO-PMO theory has had much to do with the expanding popularity of theory among experimentalists. Now, the characteristic feature of this approach is simplicity. Thus, in the vast majority of applications, the key arguments are developed on the basis of MO interaction diagrams and through utilization of low order PT, i.e., the wavefunction is treated up to first and the energy up to second order. Hoffmann and Libit[95] have recently provided a lucid discussion of qualitative high order PT in connection with a detailed theoretical analysis of substitutent effects. With the benefit of hindsight, we now know that there exist fundamental chemical problems which can only be understood at the level of high order PT and which lie beyond the range of the present day qualitative theoretical arsenal. The advantage of VB methodology is that a high order PT interpretation of VB eigenfunctions can be presented in pictorial format which can be understood by investigators from "all walks of chemistry". Specifically, high order PT interactions of VB CW's correspond to the indirect coupling of CW's which we have defined and exemplified before. In the case of the 4e-4c system, we can say that the Möbius AO array is energetically superior to the Hückel AO array because indirect coupling of each HL CW with the doubly "ionic" and the extrinsic CW's to second order (in wavefunction) is responsible for a higher order coupling of the HL CW's themselves. Actually, the indirect coupling of the HL by the doubly "ionic" CW's is due to the fact that the latter include the common CW's. Furthermore, it is worth noting that coupling via the common CW's is possible in both Hückel and Möbius AO systems, though it is more effective in the latter system. By contrast, coupling via the extrinsic CW's can occur only in the Möbius AO system.

N. Neglect of Differential Overlap Valence Bond Theory of Stereoselection

HVB theory represents the crudest form of VB theory. Nonetheless, because of the pictorial aspects of VB theory, HVB theory leads one to anticipate exactly how inclusion of classical terms which are neglected at this level of theory will modify the conclusions reached on the basis of HVB theory. Now, the performance of a complete VB analysis of a given problem is quite a cumbersome task made more undesirable by the fact that we seek "qualitative" rather than "quantitative" answers. Thus, we must identify an approximate form of VB theory which makes simple computations possible without serious loss of detail. Two candidates that come immediately to mind are the EHVB and NDO-VB theories.

Which of the two types of approximate VB theories gives a better account of electron delocalization? The question can be restated as follows: Which of the two brands of approximate theory give a better account of the relative energies of HL and "ionic" CW's and their interaction?

Consider the energy difference between the HL and "ionic" VB CW's required for the description of an electron pair bond according to VB, EHVB, and NDO-VB theories. Table 4 reveals that, if $\varepsilon_1-\varepsilon_2$ and K_{12} are arbitrarily set to zero. the VB energy gap separating Φ_2 and Φ_3 is exactly as large as the sum of the corresponding EHVB and NDO-VB energy gaps. Furthermore, if we assume that $V_a \simeq V_d$, the NDO-VB energy gap is larger than the EHVB one because the following in equality holds for most values of s:

$$J_{11} - J_{12} > - 2\beta s$$

For example, in the case of the two-electron pi bond of ethylene, we have:

$$J_{11} - J_{12} = 7.67\ \text{eV}$$

$$-2\beta s = 4.54\ \text{eV}$$

Note that, as interatomic distance increases, both s and J_{12} tend to zero and the NDO-VB energy gap becomes an increasingly better approximation of the rigorous VB energy gap. When $s = J_{12} = 0$, the NDO-VB and VB predictions became identical.

On the basis of these considerations, it is evident that NDO-VB theory is a much better approximate version of VB theory than EHVB theory. The superiority of the NDO-VB over the EHVB method is maximal when s=0 and minimal when s is large. The fact that EHMO theory has been relatively successful when applied to ground stereochemical problems is a reflection of the fact that energy differences due to interelectronic repulsions (e.g., $J_{11}-J_{12}$ in our example) become of the same order of magnitude as energy differences due to exchange interaction (e.g., $2\beta s$ in our example) at ground equilibrium geometries where interaction distances are relatively small and, as a result, s and J_{12} relatively large.

In the previous section we saw how the HVB wavefunctions are amenable to a direct qualitative interpretation which leads to a simple prediction of the energy ranking of the low energy states of Hückel and Möbius arrays. A comparison of the computed NDO-VB wavefunctions with the computed HVB wavefunctions can be made in Tables 6 and 8. It is evident that the intuitive expectations regarding the effect of electron repulsion on the energy ranking of eigenstates are all confirmed. The important contribution of NDO-VB theory is that it provides us with a set of qualitatively correct eigenfunctions for model pericyclic systems which can form the basis for the analysis of a variety of interesting chemical problems. Indeed the NDO-VB wavefunctions of Tables 6 and 8 will constitute the starting point of the discussion in many subsequent papers.

One of the great advantages of VB theory lies in the fact that the nature of the VB CW basis set permits an explicit and unequivocal definition of inter- and intraatomic nucleus-electron and electron-electron interaction. In turn, this allows for a clear definition of the so called effective one electron Hamiltonian of HMO and HVB theory. Let us see how this is done in detail.

The complete electronic Hamiltonian operator for a polyelectronic system can be written in the manner shown below, where i is an electron and M a nucleus index and the rest of the symbols have well known meaning:

$$\hat{H} = \sum_i -(1/2)\nabla_i^2 - \sum_i \sum_M Z_M/r_{iM} + \sum_{j>i} 1/r_{ij} \tag{183}$$

We can now use the following definitions, where the absence of a prime on V and J indicates that these operators act in an intraatomic sense while the presence of a prime indicates that they act in an interatomic sense.

$$\hat{K} = \sum_i -(1/2)\nabla_i^2 \tag{184}$$

$$\hat{V} + \hat{V}' = \sum_i \sum_M -Z_M/r_{iM} \tag{185}$$

$$\hat{J} + \hat{J}' = \sum_{j>i} 1/r_{ij} \tag{186}$$

The complete Hamiltonian operator now takes the form:

$$\hat{H} = \hat{K} + \hat{V} + \hat{V}' + \hat{J} + \hat{J}' \tag{187}$$

The Hückel type Hamiltonian, $\hat{H}^{o}$, can now be defined as follows:[96]

$$\hat{H}^{\circ} = \hat{K} + \hat{V} \tag{188}$$

Accordingly, the complete Hamiltonian can be written as a sum of the Hückel type Hamiltonian and a perturbation operator $\hat{P}$ defined in the manner shown below:

$$\hat{H} = \hat{H}^{\circ} + \hat{P} \tag{189}$$

$$\hat{P} = \hat{V}' + \hat{J} + \hat{J}' \tag{190}$$

Realizing that the matrix elements over $\hat{V}'$ and $\hat{J}'$ will tend to cancel and that the matrix elements over $\hat{J}$ represent the largest perturbation terms as they include self repulsion integrals, we can simplify the perturbation operator as follows:

$$\hat{P} \simeq \hat{J} \tag{191}$$

With this definition of the perturbation operator, it is evident that the first order correction of an HVB eigenstate will be positive if the eigenstate has contributions from "ionic" CW's, i.e., each HVB eigenstate will be raised in energy by an amount proportional to its percent "ionic" character. The second order energy correction will arise as a result of the interaction of the HVB states over the $\hat{P}$ operator and it may raise or lower the energy of a given state. The above constitutes a recipe for deriving the energies of NDO-VB states starting from HVB states and using perturbation theory. Higher order corrections will not be considered.

As an illustrative example, let us consider the derivation of the three NDO-VB electronic states of H_2 by a perturbative treatment of the corresponding HVB electronic states, which are defined as follows:

$$\Psi_G^o = N\,(\Phi + \Phi^+) \qquad \text{(Ground State)} \qquad (192)$$

$$\Psi_S = N'\Phi^- \qquad \text{(Singly Excited State)} \qquad (193)$$

$$\Psi_D = N''(\Phi - \Phi^+) \qquad \text{(Doubly Excited State)} \qquad (194)$$

with

$$\Phi = L(H\cdot \;\; \cdot H) \qquad (195)$$

$$\Phi^+ = L'(H^+ \;\; H:^- + H:^- \;\; H^+) \qquad (196)$$

$$\Phi^- = L''(H^+ \;\; H:^- - H:^- \;\; H^+) \qquad (197)$$

Figure 12 gives a pictorial account of perturbation due to interelectronic repulsions.

It is instructive to consider briefly how interelectronic repulsion affects the energy of a given HVB state, e.g., the ground HVB state of H_2. The energy expression is:

$$E_G = E_G^o + \Delta_G - \Delta_{GD} \qquad (198)$$

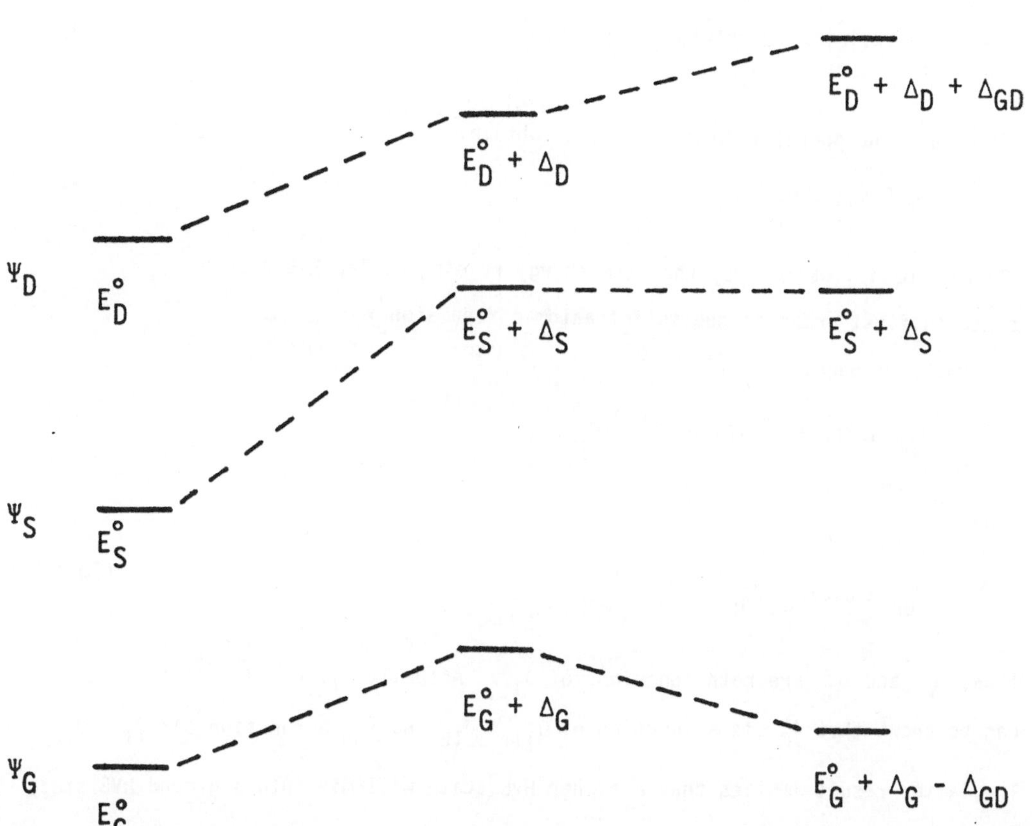

Figure 12: Perturbation of electronic states due to interelectronic repulsion.

where E_G^o is the energy of the ground HVB state, Δ_G is the first order and Δ_{GD} the second order energy correction. We can write:

$$\Delta_G \alpha <(\Phi + \Phi^+) |\hat{P}| (\Phi + \Phi^+)> \quad (199)$$

$$\alpha <\Phi|\hat{P}|\Phi> +2 <\Phi|\hat{P}|\Phi^+> + <\Phi^+|\hat{P}|\Phi^+> \quad (200)$$

If we use the operator form, $\hat{P} = \hat{J}$, we obtain:

$$\Delta_G \alpha J_{11} \quad (201)$$

On this basis, we can say that the energy raising of the HVB ground state to first order is due to intraatomic repulsions.

Next, we have:

$$\Delta_{GD} \alpha <(\Phi + \Phi^+)|\hat{P}|(\Phi - \Phi^+)>/E_G - E_D \quad (202)$$

and

$$\Delta_{GD} \alpha J_{11}/E_G - E_D \quad (203)$$

Thus, Δ_G and Δ_{GD} are both functions of J_{11}. Actually, if $\hat{P} = \hat{J} + \hat{J}' + \hat{V}'$ it can be shown that Δ_G is a function of $J_{11} + J_{12}$ and Δ_{GD} a function of $J_{11} - J_{12}$. The latter result implies that a higher HVB state will mix into a ground HVB state to an extent proportional to the difference in electron repulsions of the component CW's of the two states and inversely proportional to the energy separation of the interacting HVB states, the latter being determined by the interaction of the component CW's of the two states. That is to say, the numerator of equation (203) acts as to relieve electron repulsion at the expense of one-electron bonding through AO overlap while the denominator acts in exactly the opposite manner, precisely as expected.

On the basis of the above discussion, we obtain a highly pictorial view of how we can correct the "coulomb correlation" deficiency of the simple HVB theory and, by doing so, make a transition from HVB to approximate NDO-VB level of theory. Henceforth, we shall apply the following recipe:

a) The energy of an HVB state will be raised to first order proportionally to the percentage of "ionic" CW's which make up this state.

b) The energy of an HVB state will be lowered to second order if we can locate a low energy higher lying HVB state which can mix with it so that the contribution of the "ionic" CW's is reduced. In the case of the ethylene pi ground state this was achieved by adding Ψ_D to Ψ_G and, thus, reducing the percentage "ionic" character of the state.

0. The Electronic Basis of Stereoselection

At the HMO theoretical level, stereoselection arises whenever AO's interact (overlap) in a cyclic manner. At the HVB theoretical level, stereoselection arises whenever CW's interact (overlap) in a cyclic manner so that the product of the CW interaction matrix elements (overlap intergrals)is an odd function of the corresponding AO "resonance integrals"(overlap integrals). For the latter condition to be met, the interacting CW's must differ by one spin orbital. Accordingly, we can say that stereoselection is a natural consequence of "covalent" and/or "ionic" delocalization. That is to say, if we arbitrarily define our frame of reference to be the hypothetical situation of zero AO overlap, stereoselection arises as a result of direct or indirect mixing of higher energy "covalent" as well as "ionic" CW's with the HL CW's via one electron transfer from one AO to another. In chemistry, we frequently encounter two different types of systems wherein stereoselection arise from two different types of monoelectronic coupling of HL CW's:

a) Systems where the HL CW's are directly coupled in a one-electron sense.

b) Systems where the HL CW's are indirectly coupled in a one-electron sense via higher energy "covalent" and/or "ionic" CW's.

This classification is distinctly different from the usual classification of aromatic and antiaromatic systems according to the number of electrons or atomic centers. Thus, for example, we can find two n-electron odd annulenes one of which constitutes a type (a) and another which constitutes a type (b) case, e.g., the pi systems of cyclopropenyl anion and cyclopentadienyl cation, respectively, if non-neighbor overlap is neglected.

Let us now trace the electronic origin of stereoselection in a way which introduces schematic notation which will be particularly helpful in future applications of the theory. The necessary but not sufficient prerequisite of stereoselection is a set of cyclically interacting CW's. We can denote these CW's and their interaction by inscribing within a circle a regular polygon where the vertices represent the CW's and the sides the corresponding interaction matrix elements. For example, in the case of the 2e-3c system, we obtain the diagram shown in Figure 13a. In more complex systems, we follow an analogous procedure with the only difference that the CW's and their interaction are denoted by inscribing within a sphere a regular polyhedron where the vertices and sides have the same meaning as before. Thus, for example, in the case of the 2e-4c system, we obtain the diagram shown in Figure 13b. Note that we have made a transition from a two dimensional to a three dimensional drawing.

Next, we note that each side of a given polygon represents H_{ij} which is proportional to β_{tu}. In systems where all AO's are of the same type we can replace β_{tu} by β. If we define the number of sides of a given polygon by r, we can define the stereoselection function, ST, shown below:

$$ST = (\pm\beta_1)(\pm\beta_2)\ldots.(\pm\beta_r) \tag{204}$$

This means that stereoselection is a function of the sign of each interaction matrix element over determinantal wavefunctions. <u>That is to say, stereoselection is the necessary consequence of the antisymmetry property of the electronic wavefunction</u>.

(a)

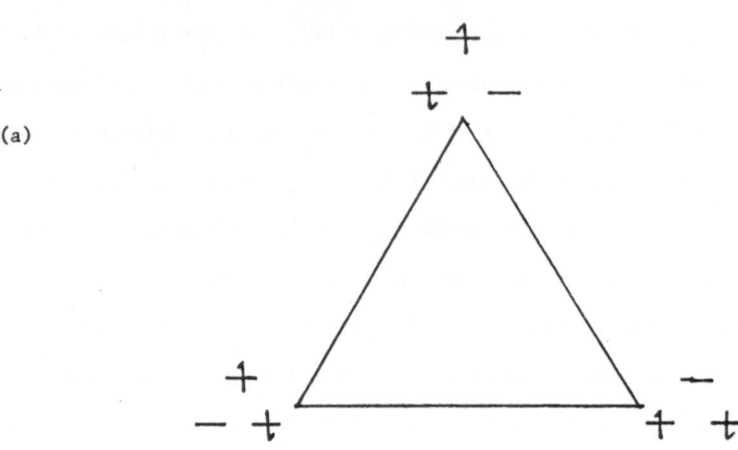

(b)

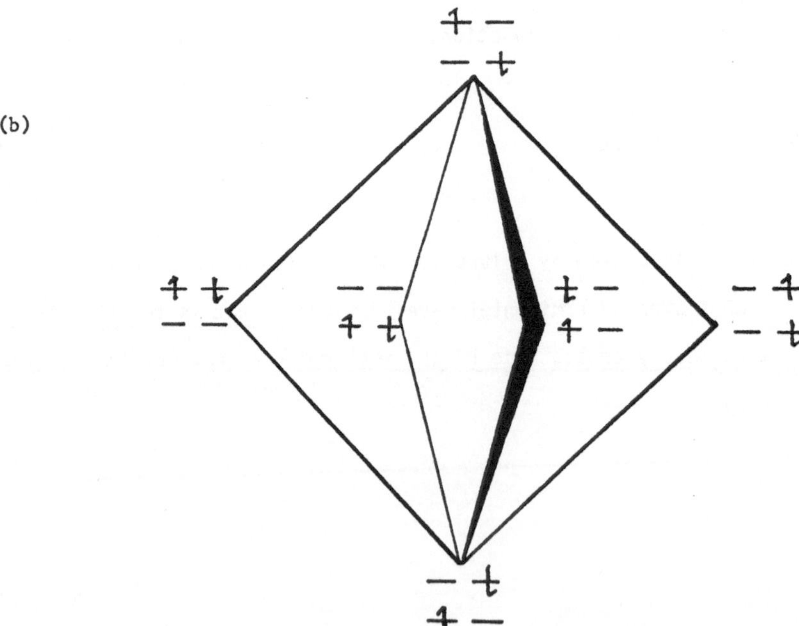

Figure 13: The mode of coupling of "covalent" CW's in two different systems:

a) 2e-3c system b) 2e-4c system

Note: The HL CW's are directly coupled in (a) but only indirectly coupled in (b) by higher energy "covalent" CW's.

P. Conclusion

This work is essentially a four part introduction to qualitative VB theory. The first part is a pedagogical exposition of the various brands of MO theory and their formal and conceptual difficulties and it constitutes the justification of this series of papers. It is followed by a discussion of VB theoretical methods and their interrelationships with MO theoretical methods culminating with the definition of equivalent VB and MO theoretical methods. The important message of this second part is that the theoretical deficiencies of a given brand of MO theory can be best understood by translating the MO theory into the equivalent VB theory and taking advantage of the fact that VB theory can project pictorially the critical interplay of "classical" and overlap effects. We shall be using this approach in exposing the limitations of the current central concepts of chemistry based on monodeterminantal MO theory. The third part of this paper has been devoted to the development of conceptual tools which can render analyses of complex problems concise and unambiguous. Thus, the matrix elements were formulated in a way which makes not only a separation of "classical" and overlap effects possible, but, in addition, exposes the energy terms contained in the simple EHMO expansions. Furthermore, the all important monoelectronic parts of matrix elements were given a diagrammatic representation which reveals immediately how overlap effects control stereoselection. Finally, a distinction of "covalent" and "ionic" delocalization was made. We shall be using these tools in many of the papers which follow in a way which will reveal clearly their potency and significance. Finally, the fourth part of this work has been devoted to an illustrative application of qualitative theory to the problem of chemical stereoselection in a manner which reveals differences and similarities, disadvantages and advantages of different brands of VB theory.

All different VB treatments of stereoselection reported in this paper confirm Hückel's rule, the Woodward-Hoffmann rules, and other equivalent stereoselection rules. Beyond this finding, we have come to recognize the following important facts:

a) Stereoselection is a function of electron delocalization and, in particular, "ionic" and "covalent" delocalization. Now, "ionic" delocalization defines electron pair bonding while "covalent" delocalization defines one- or three-electron bonding as illustrated below:

Two-Electron Bond: $A^{+}\ \ \dot{A}{:}^{-} \leftrightarrow A{\cdot\cdot}A \leftrightarrow A{:}^{-}\ \ A^{+}$

Three-Electron Bond (or, Three-Electron Antibond): $A{:}^{-}\ {\cdot}A \leftrightarrow A{\cdot}\ A{:}^{-}$

One-Electron Bond: $A^{+}\ {\cdot}A \leftrightarrow A{\cdot}\ A^{+}$

It is then apparent that aromaticity and antiaromaticity in closed shell systems are consequences of bonding modes which are traditionally studied by different branches of chemistry, namely, "even electron bonds" explored in the "classical" chemistry of ions and closed shell species and "odd electron bonds" explored in radical chemistry.

b) In even center systems having as many electrons as centers, stereoselection arises primarily from interfragmental charge transfer, i.e., an electron hop from one elementary bond to another or transfer of an electron from one fragment to the other. Such electron transfer is not energetically favorable at the level of reactants. This means that in a chemical reaction which involves some intermediate cyclic complex, stereoselection is maximally exerted in the vicinity of the transition state because of the fact that the critical CW's describing the electron transfer (e.g., Φ_{17}-Φ_{20} in Figure 10) involve three electron overlap repulsion which is expected to come strongly into play at small interfragmental distances. It follows that in even center cyclic systems of this type stereoselection is differentially exerted along the reaction coordinate. We shall discuss these matters in greater detail in a following paper dealing with the problem of the origin of reaction barriers.

c) Stereoselection is a function of electron delocalization. The greater the delocalization of a given system, the greater the "force" of stereoselection, in most cases. This suggests a strategy for "violating" the Woodward-Hoffmann rules in the following way: Consider two hypothetical "cyclobutadienes", a Hückel and Möbius cyclobutadiene, as shown below:

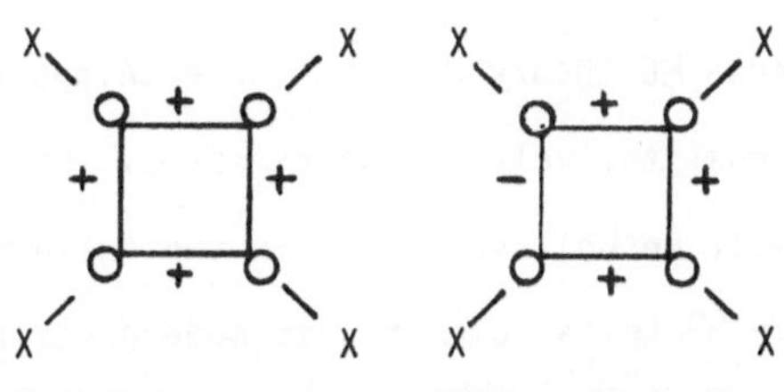

Hückel AO System Möbius AO System

We can ask the question: How do the X groups control the electron delocalization of the pi framework <u>via sigma-pi interaction?</u> An answer to this question implies a "recipe" for minimal, or, Heitler-London delocalization. With such a recipe at hand, we can design an experimental test of whether the small energetic benefit of the Möbius over the Hückel AO array in the absence of extensive delocalization is chemically meaningful. A following paper reports "violations" of the Woodward-Hoffmann rules which could have been rationally predicted using the strategy outlined above.

In recent years, there has been an explosion of interest in qualitative MO theory principally as a result of the success of the "orbital symmetry" rules of Woodward and Hoffmann, and related contributions. An ever increasing number of researchers from both the theoretical and experimental wings of chemistry have made important contributions towards the development and application of qualitative theoretical models. By contrast, very little has been done with qualitative VB theory. This is a reflection of the fact that this type of theory has a relatively high "intellectual barrier" which tends to discourage the novice. This "barrier" is due to the fact that symmetry constraints are not projected as easily and pictorially as in MO theory, matrix elements over determinantal wavefunctions are not palatable to the experimentalist who is looking for a "quick and simple" way for generating new ideas, etc. An indirect consequence of all this is the apparent impression of many chemists that VB theory is not worth pursuing because MO theory does the same things as well and in a much simpler way, at least qualitatively. The extent of the fallacy represented by such an argument will be well exhibited in the following series of papers where we show that, save Hückel's rule and its modern day reformulation, no concept developed on the basis of monodeterminantal MO theory escapes unscathed a VB theoretical reexamination. It is within the context of this neglect of VB theory that past contributions towards the generation of a qualitative VB theoretical model assume great significance and bear the indelible mark of originality. As we have mentioned before, VB theory was parented by Heitler and London. Slater, Van Vleck, Coulson, McWeeny and others showed in early papers how the VB theory of simple diatomics could be extended to polyatomic systems and how it could be formalized in an *ab initio* sense.

Pauling spearheaded the qualitative application of VB theory to large molecules of direct interest to the experimental chemists. It can be said that Pauling's monograph entitled "The Nature of the Chemical Bond" represents the conclusion of the first era of development of VB theory.

In recent times, a VB theoretical interpretation of chemical stereoselection within a class of cyclic systems was expounded in the papers of Oosterhoff and Van der Lugt and Goddard. Oosterhoff and van der Lugt[97] showed that stereoselection in Ne - Nc systems depends on high order AO overlap terms of the off-diagonal matrix elements. The more complete treatment of the same problem via the diagrammatic approach espoused in this paper (section F) is entirely equivalent to the Oosterhoff-van der Lugt approach. These authors richly deserve credit for the first recognition of the importance of high order overlap terms in the expansion of VB matrix elements. In recent times, Goddard has made use of the so called Generalized VB (GVB) method,[98] developed by him and his collegues, in order to make quantitative and qualitative predictions with regards to chemical problems spanning organic, inorganic and photochemistry. The GVB method is the quantitative form of $\overline{\text{HL}}$ theory. Goddard did not view the problem of chemical stereoselection from the vantage points of diagrammatic HL theory or conventional VB theory, as advocated in this work. Rather, he examined the shapes of the delocalized Coulson-Fischer type AO's used in connection with HL theory. In this way, he was able to develop the Orbital Phase Continuity Principle[93] which leads to the same conclusions as the alternative VB approach of Oosterhoff and van der Lugt. Finally, Salem,[99] independently as well as in marvelous collaboration with Turro and Dauben,[100] has shown that elementary, minimal basis set VB theory can be profitably applied to nonpericyclic photochemical reactions.

The VB theoretical tools developed in this paper will be used in order to analyze the structure and reactivity of model "few electrons in few orbitals" systems. We expect to uncover new fundamental concepts through a detailed and clear analysis of stereoselection and regioselection in open shell systems, spin multiplicity effects, "sigma-pi" couplings, and other problems which are presently outside the range of modern chemical concepts.[101] The point will be reached where further progress can be made only if we become capable of handling large systems. For this reason, we need to reformulate the concepts of VB theory in a way which makes large systems subject to a "back of the envelope" theoretical treatment while maintaining sufficient rigor so that our approach represents an improvement, in a formal as well as a conceptual sense, over past and present treatments. This important formulation is described in part II of this work.

REFERENCES

1. Scott, W. T. "Erwin Schrödinger"; Univ. of Mass. Press: Amherst, Mass., 1967. Phillips, J. C. Rev. Mod. Phys. 1970, 42, 317.

2. A one-electron MO perturbation theoretical approach to structural chemistry is described in: Epiotis, N.D.; Cherry, W.R.; Shaik, S.; Yates, R.L.; Bernardi, F. Top. Curr. Chem. 1977, 70, 1.

3. The Linear Combination of Fragment Configurations (LCFC) approach, essentially an approximate Molecular Orbital-Valence Bond approach, is described in:
   (a) Epiotis, N.D. Angew. Chemie, Int. Ed. Engl. 1974, 13, 751.
   (b) Epiotis, N.D.; Shaik, S. Prog. Theor. Org. Chem. 1977, 2, 348.
   (c) Epiotis, N.D.; Shaik, S. J. Am. Chem. Soc. 1978, 100, 1, and subsequent papers.
   (d) Epiotis, N.D. "Theory of Organic Reactions"; Springer-Verlag: Berlin and New York, 1978.
   (e) Epiotis, N.D. Pure and Appl. Chem. 1979, 51, 203.
   (f) Epiotis, N.D.; Shaik, S.; Zander W. in "Rearrangements in Ground and Excited States", Vol. 2, De Mayo, P., Ed.; Academic Press: New York, 1980.

4. Schaefer III, H.F. "The Electronic Structure of Atoms and Molecules"; Addison-Wesley: Reading, Massachusetts, 1972.

5. (a) Hall, M.B. Inorg. Chem. 1978, 17, 2261.
   (b) Palke, W.E.; Kirtman,B. J. Am. Chem. Soc. 1978, 100, 5717.

6. (a) Sovers, O.T.; Kern, C.W.; Pitzer, R.M.; Karplus, M. J. Chem. Phys. 1968, 49, 2592.

(b) Lowe, I.P. Science 1973, 179, 527.

(c) Brunck, T.K.; Weinhold, F. J. Am. Chem. Soc. 1979, 101, 1700.

(d) Gavezzotti, A.; Bartell, L.S. J. Am. Chem. Soc. 1979, 101, 5142.

(e) Corcoran, C.T.; Weinhold, F. J. Am. Chem. Soc. in press.

7. (a) Ref. 2, Sections 2.0-4.3 and 13.

(b) Hoffmann, R.; Levin, C.C.; Moss, R.A. J. Am. Chem. Soc. 1973, 95, 629.

(c) Nascimento, C.C.; Brinn, I.M. Z. Naturforsch 1978, 33a, 366.

8. Davydov, A.S. "Quantum Mechanics"; Pergamon Press: New York, 1965.

9. Dalgarno, A. in "Quantum Theory", Vol. 1, Bates, D.R., Ed.; Academic Press: New York, 1961.

10. (a) Hund, F. Z. Physik. 1927, 40, 742; ibid. 1927, 42, 93; ibid. 1928, 51, 759.

(b) Mulliken, R.S. Phys. Rev. 1928, 32,186; ibid. 1928, 32, 761; ibid. 1929, 33, 730.

(c) Bloch, F. Z. Physik. 1928, 52, 555.

(d) Hückel, E. Z. Physik. 1930, 60, 423.

11. (a) Heitler, W.; London, F. Z. Physik. 1927, 44, 455.

(b) Slater, J. C. Phys. Rev. 1931, 38, 1109.

(c) Van Vleck, J.; Sherman, A. Rev. Mod. Phys. 1935, 7, 168,200.

12. Methodologies for the approximate solutions of the Schrodinger equation are reviewed in: Schaefer, H.F., III, Ed. "Modern Theoretical Chemistry. Electronic Structure Theory" Vol. 3; Plenum Press: New York, 1977.

13. (a) Hückel, E. Z. Physik. 1931, 70,204; ibid. 1932, 76,628. Hückel, E. Z. Electrochem. 1937, 43,752.

(b) Streitwieser, Jr., A. "Molecular Orbital Theory for Organic Chemists"; John Wiley and Sons, Inc.: New York, 1961.

14. (a) Wolfsberg, M.; Helmholz, L. J. Chem. Phys. 1952, 20, 837.

(b) Hoffmann, R.; Lipscomb, W.N. J. Chem. Phys. 1962, 36,2189.

(c) Hoffmann, R. J. Chem. Phys. 1963, 39, 1397.

15. An excellent introduction to applied quantum chemistry and, in particular, to the "how to do it" aspects of HMO and EHMO theories can be found in McGlynn, S.P.; Vanquickenborne, L.G.; Kinoshita, M.; Carroll, D.G. "Introduction to Applied Quantum Chemistry"; Holt, Rinehart, and Winston, Inc.: New York, 1972.

16. (a) Walsh, A.D. J. Chem. Soc. 1953, 2260, 2266, 2288, 2296, 2301.

(b) Walsh, A.D. Progress in Stereochemistry 1954, 1.

(c) Mulliken, R.S. J. Am. Chem. Soc. 1955, 77, 887.

17. (a) Woodward, R.B.; Hoffmann, R. J. Am. Chem. Soc. 1965, 87, 395.

(b) Longuet-Higgins, H.C.; Abrahamson, E. J. Am. Chem. Soc. 1965, 87, 2045.

(c) Hoffmann, R.; Woodward, R.B. J. Am. Chem. Soc. 1965, 87, 2046.

18. Dewar, M.J.S. "The Molecular Orbital Theory of Organic Chemistry"; McGraw-Hill: New York, 1969.

19. The "father" of the FO approximation in "qualitative" MO theory is K. Fukui. Fukui, K.; Yonezawa, T.; Shingu, H. J. Chem. Phys. 1952, 20, 722; Fukui, K.; Yonezawa, T.; Nagata, C.; Shingu, H. J. Chem. Phys. 1954, 22, 1433.

20. For applications of the FO-PMO model to problems of structure chemistry, see ref. 2 and references therein. For applications of the FO-PMO model to problems of chemical reactivity, see:

(a) Ref. 18

(b) Hudson, R.F. Angew. Chemie, Int. Ed. Engl. 1973, 12, 36.

(c) Klopman, G. in "Chemical Reactivity and Reaction Paths", Klopman, G., Ed.; Wiley-Interscience: New York, 1974.

(d) Fukui, K. "Theory of Orientation and Stereoselection"; Springer-Verlag: Berlin and New York, 1975.

(e) Fleming, I. "Frontier Orbitals and Organic Chemical Reactions"; John Wiley: New York, 1976.

21. (a) Bartell, L.S. J. Chem. Ed. 1968, 45, 754.

(b) Pearson, R.G. "Symmetry Rules for Chemical Reactions"; Wiley and Sons, Inc.: New York, 1976.

22. (a) Pople, J. A.; Beveridge, D.L. "Approximate Molecular Orbital Theory"; McGraw-Hill: New York, 1971.

(b) Bingham, R.C.; Dewar, M.J.S.; Lo, D.H. J. Am. Chem. Soc. 1975, 97, 1285.

(c) Dewar, M.J.S.; Thiel, W. J. Am. Chem. Soc. 1977, 99, 4899.

(d) Schweig, A.; Thiel, W. J. Am. Chem. Soc. 1981, 103, 1425-1431.

23. (a) Pople, J. A., Acc. Chem. Res. 1970, 3, 217.

(b) Ref. 4.

24. These can be obtained from the Quantum Chemistry Program Exchange, Department of Chemistry, Indiana University.

25. Heilbronner, E.; Bock, H. "Das HMO-Modell und Seine Anwendung"; Verlag Chemie, Gmbh: Weinheim, 1968.

26. (a) Woodward, R.B.; Hoffmann, R. "The Conservation of Orbital Symmetry"; Verlag Chemie: Weinheim, 1970.

(b) The independent recognition of orbital symmetry control of reaction stereochemistry by a number of brilliant investigators is documented in a recent article: Epiotis, N.D.; Shaik, S.; Zander, W. in "Rearrangements in Ground and Excited States", de Mayo, P., Ed., Vol. 2; Academic Press, Inc.: New York, 1980.

27. The "father" of the concept of aromaticity is E. Hückel (see ref. 13a). Its applicability to problems of chemical reactivity and, in particular, pericyclic reactions was recognized independently, under different theoretical disguises, by M. G. Evans, M. J. S. Dewar, and H. E. Zimmerman:

(a) Evans, M.G. Trans. Faraday Soc. 1939, 35, 824.

(b) Dewar, M.J.S. Angew. Chem., Int. Ed. Engl. 1971, 10, 761.

(c) Zimmerman, H.E. Acc. Chem. Res. 1971, 4, 272.

28. (a) Wheland, G.W. J. Chem. Phys. 1934, 2, 474.

(b) Pauling, L.; Springall, H.S.; Palmer, K.J. J. Am. Chem. Soc. 1939, 61, 927.

(c) Mulliken, R.S. J. Chem. Phys. 1939, 7,339.

(d) Mulliken, R.S.,; Rieke, C.A.; Brown, W.G. J. Am. Chem. Soc. 1941, 63, 41.

(e) Dewar, M.J.S. "Hyperconjugation", Ronald Press Co.: New York, 1962.

29. The formal limitations of HMO theory are discussed in a number of treatises of quantum chemistry. A concise comparative discussion of MO methods can be found in: Richards, W.G.; Horsley, T.A. "Ab Initio Molecular Orbital Calculations for Chemists"; Clarendon Press: Oxford, 1970.

30. Equations (1)-(3) lead to the correct explicit forms of the HMO energy expressions. Actually, equation (3) holds only for the diagonal but not the off diagonal elements of the HMO energy matrix which are parametrized so that effectively the integral of equation (3) is nonzero.

31. For review of "Molecules in Molecules" theoretical approaches, see: Fabian, J. J. Signal AM 6 1978, 4, 307; Fabian, J. J. Signal AM 7 1979, 1, 67.

32. Hoffmann, R. Acc. Chem. Res. 1971, 4, 1.

33. Inagaki, S.; Fujimoto, H.; Fukui, K. J. Am. Chem. Soc. 1976, 98, 4693.

34. Levin, A.A. "Solid State Quantum Chemistry", McGraw-Hill: New York, 1977.

35. (a) Two electron-two orbital diradicals: Salem, L.; Rowland, C. Angew. Chem., Int. Ed. Engl. 1972, 11, 92 and references therein.
(b) Oxygen (O_2): Kasha, M.; Brabham, D.E. in "Singlet Oxygen", Wasserman, H.H.; Murray, R.W., Eds.; Academic Press: New York, 1979.
(c) Methylene (CH_2): Borden, W.T.; Davidson, E.R. Ann. Rev. Phys. Chem. 1979, 30, 125 and references cited therein.
(d) Cyclobutadiene (C_4H_4): Ref. 35c and references cited therein.
(e) Trimethylenemethane (C_4H_6): Ref. 35c and references therein.
(f) Benzyne (C_6H_4): Wilhite, D.L.; Whitten, J.L. J. Am. Chem. Soc. 1971, 93, 2858.

36. (a) Hosteny, R.P.; Dunning, T.H., Jr.; Gilman, R.R.; Pipano, A.; Shavitt, I. J. Chem. Phys. 1975, 62, 4764.
(b) For other recent theoretical investigations, see: Buenker, R.J.; Shih, S.; Peyerimhoff, S.D. Chem. Phys. Lett. 1976, 44, 385. Schulten, K.; Ohmine, I.; Karplus, M. J. Chem. Phys. 1976, 64, 4422.
(c) Electron impact study: Doering, J.P. J. Chem. Phys. 1979, 70, 3902.

37. Hudson, B.S.; Kohler, B.E. Chem. Phys. Lett. 1972, 14, 299. Hudson, B.S.; Kohler, B.E. J. Chem. Phys. 1973, 59, 4984.

38. Dixon, D.A.; Stevens, R.M.; Herschbach, D.R. Faraday Disc. Chem. Soc. 1977, 62, 110.

39. Keil, F.; Ahlrichs, R. J. Am. Chem. Soc. 1976, 98, 4737.

40. For a recent review of the Diels-Alder reaction, see: Sauer, J.; Sustmann, R. Angew. Chem., Int. Ed. Engl. 1980, 19, 779.

41. Dewar, M.J.S.; Olivella, S.; Rzepa, H.S. J. Am. Chem. Soc. 1978, 100, 5650.

42. (a) For review of polyfluoroalkene and polyfluorodiene cycloadditions, see: Chambers, R.D. "Fluorine in Organic Chemistry"; John Wiley and Sons: New York, 1973, pp. 179-189.

43. For recent explorations of the mechanism of the S_N2 reaction on carbon, see: Gray, R.W.; Chaple, C.B.; Vergnani, T.; Dreiding, A.S.; Liesner, M.; Seebach, D. Helv. Chim. Acta 1975, 58, 2524, and references therein.

44. For reviews of the S_N2 reaction on silicon, see:

(a) Sommer, L.H. "Stereochemistry, Mechanism, and Silicon"; McGraw-Hill: New York, 1965.

(b) Corriu, R.; Masse, J. J. Organomet. Chem. 1972, 35, 51.

45. For the most recent "qualitative" theoretical analysis of the stereochemistry of the S_N2 reaction and a compilation of references to previous theoretical work on this subject, see: Anh, N.T.; Minot, C. J. Am. Chem. Soc. 1980, 102, 103.

46. The rules for the conservation of "correlation energy" have been expressed as follows:

1. The number of electron pairs must be conserved.
2. The spatial arrangement must be approximately maintained for electron pairs which are nearest neighbours.

See: Kutzelnigg, W. Topics Curr. Chem. 1973, 41, 31. In most ground stereochemical problems (e.g., conformational isomerism) these two requirements are satisfied. For example, SCF-MO calculations of internal rotation in ethane are in excellent agreement with experiment: Ref. 4.

47. The cancellation of "correlation effects" in some chemical processes has been noted early in the following works:

(a) Nesbet, R.K. J. Chem. Phys. 1962, 36, 1518.

(b) McLean, A.D. J. Chem. Phys. 1963, 39, 2653.

(c) Nesbet, R.K. Advan. Chem. Phys. 1965, 9, 321.

48. Early recognition of the stereochemical dependence of hyperconjugation: Altona, C.; Romers, C.; Havinga, E. Tetrahedron Lett. 1959, 16.

49. For recent important MO theoretical contributions towards an understanding of the stereochemical dependence of hyperconjugation, see:

(a) Radom, L.; Hehre, W.J.; Pople, J.A. J. Am. Chem. Soc. 1972, 94, 2371.

(b) Hoffmann, R.; Radom, L.; Pople, J.A.; Schleyer, P.v.R.; Hehre, W.J.; Salem, L. J. Am. Chem. Soc. 1972, 94, 6221.

(c) Ref. 2, part IV.

(d) Ref. 6c.

50. Redington, R.L.; Olson, W.B.; Cross, P.C. J. Chem. Phys. 1962, 36, 1311.

51. Jackson, R.H. J. Chem. Soc. 1962, 84, 4585.

52. (a) Yamaguchi, A.; Ichishima, I.; Shimanouchi, T.; Mizushima, S.I. Spectrochim. Acta 1960, 16, 1471.

(b) Kasuya, T.; Kojima, T. J. Phys. Soc. Japan 1963, 18, 364.

53. (a) Koster, D.F.; Miller, F.A. Spectrochim. Acta 1968, 24A, 1487.

(b) Durig, J.R.; Clark, J.W. J. Chem. Phys. 1968, 48, 3216.

(c) Johnson, F.A.; Aycock, B.F.; Haney, C.; Colburn, C.B. J. Mod. Spect. 1969, 31, 66.

(d) Colburn, C.B.; Johnson, F.A.; Haney, C. J. Chem. Phys. 1969, 43, 4526.

(e) Cardillo, M.J.; Bauer, S.H. Inorg. Chem. 1969, 8, 2086.

54. (a) P_2H_4: Elbel, S.; Dieck, H.; Becker, G; Eusslin, W. Inorg. Chem. 1976, 15, 1235.

(b) P_2F_4: Rudolph, R.W.; Taylor, R.C.; Parry, R.W. J. Am. Chem. Soc. 1966, 88, 3729.

55. For review of conformational isomerism, see: Yokozeki, A.; Bauer, S.H. Top. Curr. Chem. 1974, 53, 71.

56. Huheey, T.E. "Inorganic Chemistry"; Harper and Row: New York, 1972.

57. Schaefer III, H.F. Acc. Chem. Res. 1979, 12, 288.

58. Bastiansen, O.; Kveseth, K.; Mallendal, H. Top. Curr. Chem. 1979, 81, 99, especially pp. 116-119.

59. Some representative early work:

(a) Apeloig, Y.; Schleyer, P.v.R.; Binkley, J.S.; Pople, J.A. J. Am. Chem. Soc. 1976, 98, 4332.

(b) Collins, T.B.; Dill, J.D.; Jemmis, E.D.; Apeloig, Y.; Schleyer, P.v.R.; Seeger, R.; Pople, J.A.; ibid. 1976, 98, 5419.

(c) Jemmis, E.D.; Poppinger, D.; Schleyer, P.v.R.; Pople, J.A., ibid. 1977, 99, 5796.

60. Apeloig, Y.; Schleyer, P.v.R.; Binkley, J.S.; Pople, J.A.; Jorgensen, W.L. Tetrahedron Lett. 1976, 3923.

61. Lowe, J.P. J. Am. Chem. Soc. 1974, 96, 3759.

62. For review, see: Buenker, R.J.,: Peyerimhoff, S.D. Chem. Revs. 1974, 74, 127.

63. (a) Pople, J.A. Proc. Roy. Soc. Ser. A. 1955, 233, 233,

(b) Amos, A.T.; Musher, J.I. Mol. Phys. 1967, 13, 509.

64. (a) Sustmann, R.; Binsch, G. Mol. Phys. 1971, 20, 1, 9.

(b) Klopman, G.; Hudson, R.F. Theor. Chim. Acta 1967, 8, 165.

65. Davidson, R.B.; Allen, L.C. J. Chem. Phys. 1971, 54, 2828.

66. Eilers, J.E.; Liberles, A. J. Am. Chem. Soc. 1975, 97, 4183.

67. Kitaura, K.; Morokuma, K. Int. J. Quantum Chem. 1976, 10, 325.

68. Nagase, S.; Morokuma, K. J. Am. Chem. Soc. 1978, 100, 1661, 1666.

69. Kollman, P.A. Acc. Chem. Res. 1977, 10, 365.

70. Slater, J.C. "Quantum Theory of Molecules and Solids", Vol. I ; McGraw-Hill, Inc.: New York, 1963.

71. The manner in which CI corrects the deficiency of the monodeterminantal MO wavefunction is often illustrated by reference to the simple example of a two electron-two orbital system, e.g., H_2, pi ethylene, etc. Elementary discussions of this type can be found in a number of elementary texts. See, inter alia: Borden, W.T. "Modern Molecular Orbital Theory for Organic Chemists"; Prentice-Hall: Englewood Cliffs, New Jersey, 1975.

72. (a) Bent, H.A. Chem. Revs. 1968, 68, 587.
    (b) Schnuelle, G.W.; Parr, R. G. J. Am. Chem. Soc. 1972, 94, 8974.

73. (a) Many Body Perturbation Theory: Paldus, J.; Cizek, J. Advan. Quant. Chem. 1975, 9, 105 and references therein. Pople, J.A.; Binkley, J.S.; Seeger, R. Intern. J. Quantum. Chem. 1976, 510, 1.
    (b) Cluster Expansions: Sinanoglou, O. J. Chem. Phys. 1962, 36, 706.
    (c) Second Order Bethe-Goldstone Method: Nesbet, R.K. Adv. Chem. Phys. 1969, 14, 1.
    (d) Independent Electron Pair Approximation: Ahlrichs, R.; Lischka, H.; Staemmler, V.; Kutzelnigg, W. J. Chem. Phys. 1975, 12, 1225 and references therein [Note: This method is related to those of (b) and (c) above].
    (e) Coupled Electron Pair Approximation: Meyer, W. Intern. J. Quantum. Chem. 1971, 55, 341.

(f) "Traditional" Configuration Interaction: Shavitt, I. in "Modern Theoretical Chemistry", Schaefer, H.F., Ed.; Plenum Press: New York, 1976.

(g) Generalized Valence Bond Theory: Bobrowicz, F.W.; Goddard III, W.A. in "Modern Theoretical Chemistry", Schaefer, H.F., Ed.; Plenum Press: New York, 1976.

74. Basis set effects may be larger than "correlation effects" in many cases. For example, see:

(a) Hurley, A.C. Advan. Quantum. Chem. 1973, 7, 315.

(b) Hariharan, P.C.; Pople, J.A. Theoret. Chim. Acta 1973, 28, 213.

(c) Ref. 4

75. This and other approximations are discussed in refs. 14(a) and 15.

76. Pauling, L. "The Nature of the Chemical Bond"; Cornell University Press: Ithaca, New York, 1962.

77. Mulliken, R.S. J. Chem. Phys. 1955, 23, 1833.

78. (a) Coulson, C.A.; Rushbrooke, G.S, Proc. Camb. Phil. Soc. 1940, 36, 193.

(b) Coulson, C.A.; Longuet-Higgins, H.C. Proc. Roy. Soc. 1947, A192, 16.

(c) Coulson, C.A. "Valence", 2nd Ed.; Oxford University Press: Oxford, 1961.

79. Longuet-Higgins, H.C. J. Chem. Phys. 1950, 18, 265, 275.

80. (a) Dewar, M.J.S., "The Electronic Theory of Organic Chemistry"; Oxford University Press: Oxford, 1949.

(b) Pullman, B.; Pullman, A. "Les Theories Electroniques de la Chimie Organique"; Massae et cie: Paris, 1952.

(c) Daudel, R.; Lefebvre, R.; Moser, C. "Quantum Chemistry"; Interscience: New York, 1959.

(d) Parr, R.G. "The Quantum Theory of Molecular Electronic Structure"; Benjamin: New York, 1963.

81. Fukui, K. "Molecular Orbitals in Chemistry, Physics and Biology", Lowdin, P.; Pullman, B., Eds.; Academic Press: New York, 1964.

82. Roberts, J.D. "Notes on Molecular Orbital Calculatons", W.A. Benjamin: New York, 1962.

83. Wheland, G.W. "Resonance in Organic Chemistry"; John Wiley and Sons: New York, 1955.

84. (a) Eyring, H.; Kimball, G.E. J. Chem. Phys. 1933, 1, 239, 626.

(b) Eyring, H.; Frost, A.A.; Turkevich, J. J. Chem. Phys. 1933, 1, 777.

85. Pauling, L. J. Chem. Phys. 1933, 1, 280.

86. McWeeny, R. Proc. Roy. Soc. A. 1953, 233, 63; ibid. 1953, 223, 306.

87. (a) Pauling, L.; Wilson, E.B. "Introduction to Quantum Mechanics", McGraw-Hill Book Co., Inc.: New York, 1935.

(b) Glasstone, S.; Laidler, K.J.; Eyring, H. "The Theory of Rate Processes"; McGraw-Hill: New York, 1941.

(c) Eyring, H.; Walter, J.; Kimball, G.E. "Quantum Chemistry"; Wiley and Sons: New York, 1954.

(d) Slater, J.C. "Quantum Theory of Molecules and Solids", Vol. 1; McGraw-Hill: New York, 1963.

(e) Sandorfy, C. "Electronic Spectra and Quantum Chemistry"; Prentice-Hall: Englewood Cliffs, NJ, 1964.

88. The reader is also directed to the works of W.A. Goddard III for discussion of the physical and chemical interpretation of VB matrix elements.

89. These are the perturbation theoretical expressions for the second order correction of energy and they are discussed in a number of elementary texts of quantum chemistry.

90. Maier, G. Angew. Chem., Int. Ed. Eng. 1974, 13, 425.

91. (a) Ref. 13.

(b) Ref. 82.

92. Coulson, C.A.; Fischer, I. Phil. Mag. 1949, 40, 386.

93. Goddard III, W.A. J. Am. Chem. Soc. 1972, 94, 793.

94. The first reformulation of traditional VB theory over nonorthogonal AO's in terms of an orthogonal AO basis is due to McWeeny, who, in, fact, applied VB methodology to organic molecules: McWeeny, R. Proc. Roy. Soc. 1954, 288.

95. Libit, L.; Hoffmann, R. J. Am. Chem. Soc. 1974, 96, 1370.

96. Actually, the diagonal matrix elements is HMO theory are evaluated with respect to $\hat{K}+\hat{V}$ but the off-diagonal matrix elements are effectively evaluted with respect to $\hat{K}+\hat{V}+\hat{V}'$. However, the qualitative perturbation formulation described here does not depend on whether we choose one or the other operator representation. See also footnote 30.

97. Mulder, J.J.C.; Oosterhoff, L.J. Chem. Commun. 1970, 305, 307.

98. Goddard III, W.A.; Dunning, Jr., T.H.; Hunt, W.J.; Hay, P.J. Acc. Chem. Res. 1973, 6, 368.

99. Salem, L. Science 1976, 191, 822.

100. Dauben, W.G.; Salem, L.; Turro, N.J. Acc. Chem. Res. 1975, 8, 41.

101. Beyond and above chemistry as narrowly defined, we recognize that VB theory can be used to analyze systems which are made up by fermions, electronic systems being merely a special case. For example, the VB CW's of pi allyl can be thought of as representative of elementary particle wavefunctions (baryon octet). Indeed, VB theory has the potential of ultimately making a unified theory of matter a reality. In this connection, see footnote 5 in: Matsen, F.A., Acc. Chem. Res. 1978, 11, 387.

	Term	Energy Expression
4e-2c System		$2\varepsilon s^2 + 2\beta s$
		$4\beta s^3$
2e-3c System	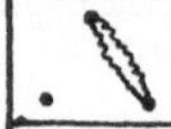	$2\beta s$
		$\varepsilon s + \beta$
		$2\beta s$
4e-3c System		$2\varepsilon s^2 + 2\beta s$
		$2\varepsilon s^2 + 2\beta s$
		$\varepsilon s^3 + 3\beta s^2$
		$4\beta s^3$
		$3\varepsilon s + \beta$
		$2\varepsilon s^2 + 2\beta s$
		$\varepsilon s^3 + 3\beta s^2$
		$\varepsilon s^3 + 3\beta s^2$
		$4\beta s^4$
4e-4c System	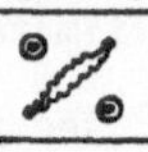	$2\varepsilon s_D^2 + 2\beta_D s_D$
		$2\varepsilon s^2 + 2\beta s$
		$\varepsilon s_D s^2 + 2\beta s s_D + \beta_D s^2$
		$4\beta s^3$
		$4\beta s^3$

Appendix 1 Energy Expressions for Monoelectronic Exchange Terms for various systems (s = nearest neighbor overlap, s_D = diagonal overlap in 4-center system).

Appendix 2. NDO-VB Computations

The NDO-VB calculations reported in the text involve an empirical parametrization of the CW energy matrix elements. Specifically, the off-diagonal elements have been computed assuming the AO "resonance integral" equals 4.65 eV. Furthermore, the relative energies of the diagonal elements have been computed by assigning zero to the HL CW's, $J_{11} - J_{12}$ to the singly "ionic" CW's, and $2J_{11} - 2J_{12}$ to the doubly "ionic" CW's using the following integral values: $J_{11} = 16.9$ eV, $J_{12} = 9.0$ eV, and for the diagonal interaction in the 4c-4e system, $J'_{12} = 6.8$ eV.

PART II

QUALITATIVE MOLECULAR ORBITAL-VALENCE BOND THEORY

Introduction

In Part I, we have shown that Valence Bond (VB) theory leads to a clear and detailed understanding of aromaticity and antiaromaticity in model systems containing no more than four electrons, four AO's, and four centers. In following papers, we shall see that VB theory can generate novel insights when applied to diverse problems with the stipulation that the target of the investigation is reduced to essential components, i.e., the target system is replaced by a model target system containing only a few electrons, a few AO's, and a few centers. The theoretical treatment of model rather than real systems is the historical compromise one makes in order to be able to extract concepts of general applicability which otherwise might have been obscured by the complexity of the problem. Today, the conceptual framework of chemistry owes much of its existence to model studies, theoretical and experimental. Indeed, much of the progress made towards an understanding of molecular structure and reactivity is due to two major approximations:

1) The sigma-pi separation approximation,[1] according to which the pi and sigma frameworks of a given system are assumed not to interact. Thus, in seeking to understand the electronic basis of antiaromaticity in cyclobutadiene and aromaticity in benzene, we focus attention on the pi array of AO's and electrons while neglecting all other sigma AO's and electrons.

2) The Frontier Orbital (FO) approximation,[2] according to which only the Highest Occupied MO's (HOMO's) and the Lowest Unoccupied MO's (LUMO's) of two interacting fragments and the electrons which they contain are taken into consideration in developing a theoretical treatment. For example, the transition state complex of ethylene dimerization is viewed as a composite of two ethylene fragments. The FO's of ethylene are the π and π * MO's, and the treatment of the transition state is based on consideration of the four FO's and the four electrons which occupy them.[3] Similarly, the gauche and anti conformational isomers of CH_2F-CH_2F are viewed as composites of two -CH_2F fragments

having different spatial orientations. The HOMO of $-CH_2F$ is mostly localized in the two C-H sigma bonds and the LUMO of $-CH_2F$ is mostly localized on the C-F sigma bond. In a model treatment, only the interaction of the C-H bonds of one fragment with the C-F bond of the second fragment is considered while the interaction of all other bond pairs is neglected.[4] Many other examples of the utilization of the FO approximation in problems of chemical structure and reactivity can be given. The interested reader is referred to the recent monograph on the theory of stereoselection by the originator of the FO approximation[5] as well as our recent monograph on the structural theory of organic chemistry (MO theory) where we make extensive use of the FO approximation.[4]

"Qualitative" theoretical treatments of model systems have been extraordinarily useful in the past and, no doubt, will continue to be so in the near future. However, the following facts suggest that development of qualitative theories of real systems is of essential importance:

a) In the last twenty years, theoretical techniques originally developed in different areas of natural science (e.g., nuclear physics) have had a great impact on theoretical chemistry. This, coupled with technological advances in computer science, has made good quality _ab initio_ computations of modestly large polyatomic systems possible. By "good quality computations", we imply calculations which employ large orbital and configuration bases and which include geometry optimization. Indeed, experimentalists have already begun to use quantum chemical computer programs as analytical tools alongside the conventional experimental probes of chemical structure and reactivity. Thus, if qualitative theory is to be an integral complement of quantitative theory and the basis of every day chemical communication and "chemical plotting", it must not only be able to handle "real", in addition to model, systems, but, it must do so at the level of "state of art" ab initio quantum theory.

b) In chemistry, as in every other science, we operate from a set of assumptions and concepts which frequently break down. When this occurs, we seek to develop new ideas of broader applicability. However, our attempts to do so are often misdirected in that the tools that we apply to the problem at hand are different yet equivalent to those used before. For example, suppose that we are interested in reaction stereochemistry and we are aware of a thermal cyclo-addition of a diene and an olefin which proceeds nominally in a $2\pi + 2\pi$ manner, i.e., in a manner opposite to the one expected on the basis of the standard Woodward-Hoffmann rules,[6] the Dewar-Zimmerman Perturbation MO (PMO) treatment,[7] and other related approaches.[8] In attempting to understand the origin of the discrepancy we can consider the following options:

1. Surmise that the original model theoretical treatment is wrong and devise a different model theoretical approach.

2. Surmise that the original model theoretical treatment is right but that sufficient difference exists between "model" and "real" to warrant the development of a more complete theoretical approach applicable to real systems which can lead to a broader outlook. In other words, we must ask the following two questions:

i. Are the orbital symmetry rules for the prediction of ground state reaction stereochemistry theoretically correct?

ii. If so, why are they breaking down when no apparent reason exists? Why are there "illegitimate" exceptions to the orbital symmetry rules?[9]

The answer to the first question can be obtained by applying different theoretical methods to model systems and comparing the conclusions with the original conclusions of Woodward and Hoffmann[6] and others.[7,8] Note that unless the theoretical treatment is sufficiently detailed and incisive, in a mathematical sense, there is no assurance that an ultimate answer of "yes" or "no" can be be given. In Part I, we saw that a VB treatment of aromaticity and antiaromaticity,[10] which incorporates all important electronic effects in an explicit manner, reveals that there is no way that the traditional orbital symmetry rules developed from consideration of model systems can be violated because stereoselection is a consequence of the determinantal nature of the total electron wavefunction.[11] This suggests that any further preoccupation with orbital symmetry problems viewed from the vantage point of model theory is an excercise in futility. To put it crudely, if we wish to understand why our hypothetical diene and olefin became thermally united in a $2\pi + 2\pi$ fashion we must not think of the cycloaddition process as "six electrons in six orbitals" but rather as "many electrons in many orbitals". If we want to go beyond current ideas, we must be able to perform what has not yet been performed, i.e., we must replace the theory of model systems by a theory of real systems.

A qualitative VB theory of polyelectronic systems which leads to the formulation of concepts which are not intuitively obvious is next to impossible, for reasons made clear in the previous paper. On the other hand, the formal and conceptual advantages of VB theory are too precious to be abandoned. Hence, we must retain the basic VB formalism but constrain the orbital basis so that a more "economic" VB theory results which can be applied to any system of interest. MOVB theory is precisely this type of "economic" VB theory.

The "economy" aspect of MOVB theory can be illustrated by reference to a specific example. Consider a four electron-three orbital-three center linear system with one AO per center such as the one shown below:

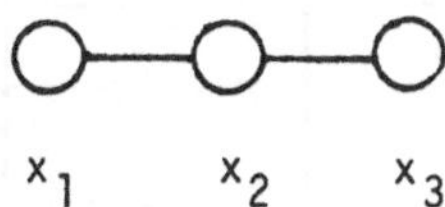

At the level of VB theory our orbital basis set is comprised of the three AO's x_1,x_2, and x_3. At the level of MOVB theory, we replace any two AO's by two MO's generated by combining the two chosen AO's in an in-phase and an out-of-phase manner. In our example, the two MO's are:

$$\omega_1 = N\ (x_1 + x_3) \tag{1}$$

$$\omega_2 = N'\ (x_1 - x_3) \tag{2}$$

Hence, the orbital basis set is now comprised of two MO's, ω_1 and ω_2, and one AO, x_2. The resulting VB and MOVB Configuration Wavefunctions (CW's) as well as their interrelationships are depicted in Table 1. Each MOVB CW, Φ_i, can be expanded into a linear combination of VB CW's, X_i, in the way implied by the cross-hatched squares. For example, Φ_2 is a linear combination of X_1 and X_2, etc. Now, in order to determine the six VB eigenstates one must solve a 6x6 determinantal equation since each single VB CW interacts with at least one other. By contrast, in order to determine the six MOVB eigenstates one must solve a 4X4 and a 2X2 determinantal equation, since the MOVB CW's Φ_1, Φ_3, Φ_4, Φ_5 belong to one symmetry type while the MOVB CW's Φ_2, Φ_6 belong to another symmetry type. In fact, if we neglect bielectronic terms and we are interested in determining the ground eigenstate only, we must still solve a 6X6 determinantal equation at the VB level but only a 3X3 one (over Φ_3, Φ_4, Φ_5) at the MOVB

Table 1. MOVB (Φ_i) and VB (X_i)CW's and the Expansion of MOVB into VB CW's*.

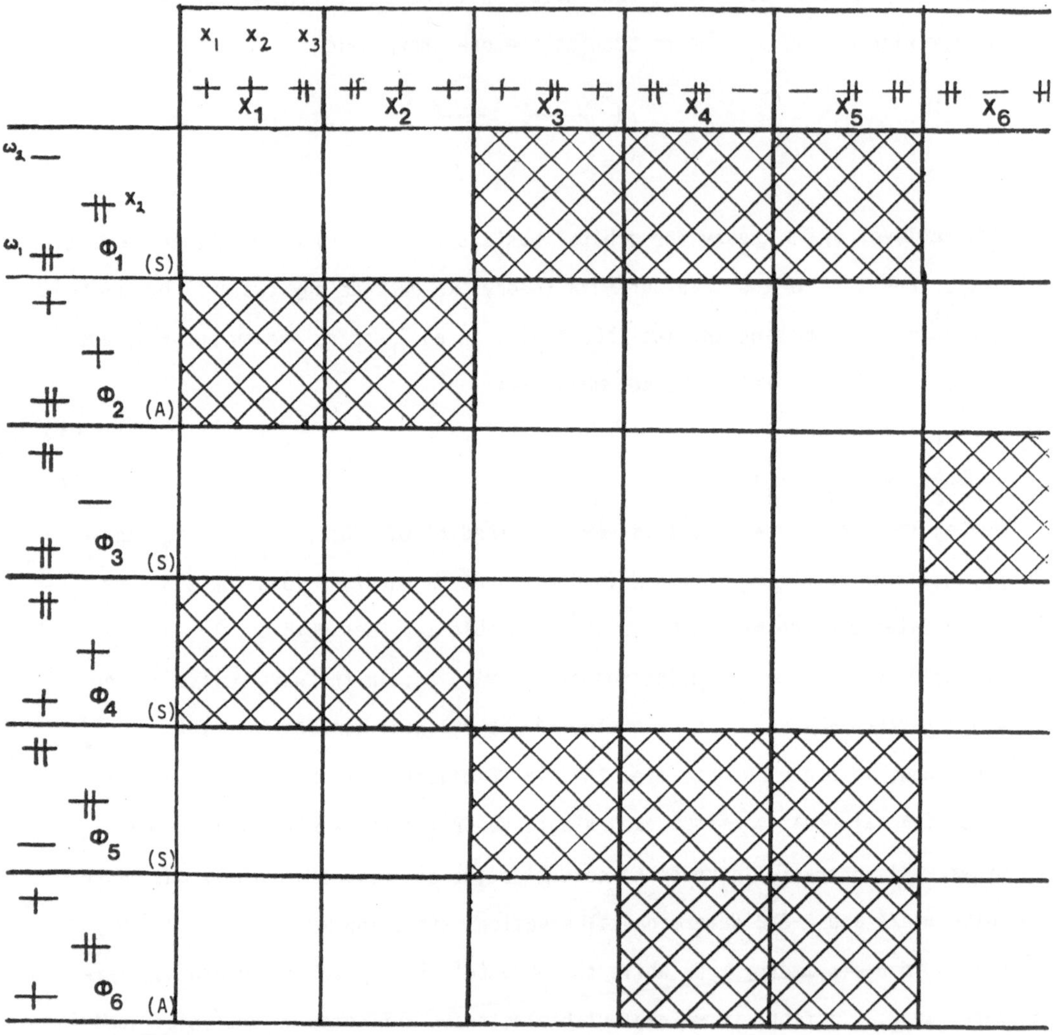

	X_1	X_2	X_3	X_4	X_5	X_6
Φ_1 (S)						
Φ_2 (A)						
Φ_3 (S)						
Φ_4 (S)						
Φ_5 (S)						
Φ_6 (A)						

* S and A are the symmetry species of each Φ_i for a symmetrical linear arrangement of the three AO's.

level. Clearly, if we desire to understand ground state chemical bonding, it is much more convenient to generate the concepts by manipulating three rather than six primitive CW's and the same is true for any other eigenstate.

Five major qualitative applications of MOVB theory to important chemical problems stand out. The very first application was proposed by Mulliken in connection with his investigations of the electronic structure of ground and excited Charge Transfer (CT) complexes.[12] The distinction between an excited CT complex and an "exciplex" is a fuzzy one but there is little doubt that the concept of the "excimer" advanced by Förster[13a] on the basis of MOVB theory has had a profound impact on photochemistry.[13c] The related studies of Weller[13b] on photochemical electron transfer also deserve special mention. A brilliant analysis of the structure of organometallic complexes using the MOVB approach has been contributed by Mason and McWeeny.[14a-c] The recently proposed "spin-pairing" model of Drago is a direct relative of the Mason-McWeeny model.[14d] Finally, stereochemical selection rules for thermal and photochemical reactions derived via MO-VB theory have been advanced by Fukui[15,5] and Epiotis[16] and a qualitative MOVB theory of reaction potential energy surfaces has been recently published.[17] All of the above treatments make use of one or more of the following approximations:

1. The Neglect of Differential Overlap (NDO) approximation.

2. The FO approximation, i.e., these treatments deal with model systems containing "few electrons in few orbitals".

3. Utilization of low order Perturbation Theory (PT) for the purpose of construction of the MOVB eigenstates.

These approximations are reasonable when the fragments interact in a weak manner, but they become totally unacceptable in the case of strong fragment interaction. Since we are interested in the development of a general theory of chemical bonding applicable to any chemical system, we must reject these three key approximations and seek to develop a new conceptual framework free from any biases and preconceptions arising from our previous entanglement with qualitative MOVB (and MO) theory.

A. MOVB Theory

The most useful physical model which can form the basis for a theoretical analysis of a chemical problem is the "Fragments in Molecules" (FIM) model. Variants of the FIM model include the "Atoms in Molecules" (AIM),[18] "Diatomics in Molecules" (DIM),[19] and "Molecules in Molecules" (MIM)[20] models. In principle, these can be used in conjunction with either MO or VB theory. The MOVB theory of chemical bonding we are about to develop makes use of a different type of FIM model, namely, the Core-Ligand (CL) model which projects how bonds are formed between an atomic or molecular core and a set of ligands. For example, we shall view methane as the product of the union of C and H_4, water as the result of combining O and H_2, etc. Each of the two fragments has an associated orbital manifold and configurations can be built by distributing all valence electrons among all valence orbitals of the two fragments in all possible ways with each MOVB CW written as a Slater determinant or a linear combination of Slater determinants. Now, in Part I, we stated that one of the disadvantages of MO theoretical models is the absence of a unique or universally acceptable reference frame. This difficulty is removed at the level of VB theory where the free atoms constitute an unambiguous frame of reference. By contrast, the problem persists in any formalism which contracts an AO basis to an MO basis, i.e., it persists at the level of MOVB theory. The great advantage of the CL model is that it copes with the "frame of reference difficulty" in the best possible way because every conceivable chemical system can be formulated as a composite of a central core and surrounding ligands, a reference electronic configuration can be unambiguously and universally defined (*vide infra*), and the bonds linking the core and the ligands can be described correctly by the MOVB method to be used in connection with this model.

Let us now consider a prototypical system of two fragments C and L with the former having two and the latter one orbital wherein a total of four electrons are confined. Six possible MOVB CW's can be generated by permuting the four electrons among the three orbitals and these are shown in Figure 1. The energy of each CW, Φ_i, is given by the expression below, in a manner entirely analogous to that of VB theory.

$$H_{ii} = E_i = F_i + G_i + X_i \tag{3}$$

or

$$H_{ii} = E_i = P_i + G_i + X_i \tag{4}$$

F_i is the (adjusted) energy of the isolated fragments. P_i represents the excitation energy of a MOVB CW and its physical significance is apparent: as P_i becomes increasingly positive, the contribution of the corresponding MOVB CW to the ground eigenstate decreases while its contribution to high energy eigenstates increases, other things being equal. Furthermore, P_i provides a basis for a systematic classification of the CW's according to excitation type. Thus, we may identify a ground CW (P_i=0), singly excited CW's resulting from a single electron transfer with respect to the ground CW, etc. Now, since electron transfer can occur from one orbital of one fragment to another orthogonal orbital of the same fragment or to another nonorthogonal orbital of another fragment, we can distinguish between intrafragmental excitation, commonly referred to as <u>local excitation</u>, and interfragmental excitation, commly referred to as <u>Charge Transfer</u> (CT) <u>excitation</u>.

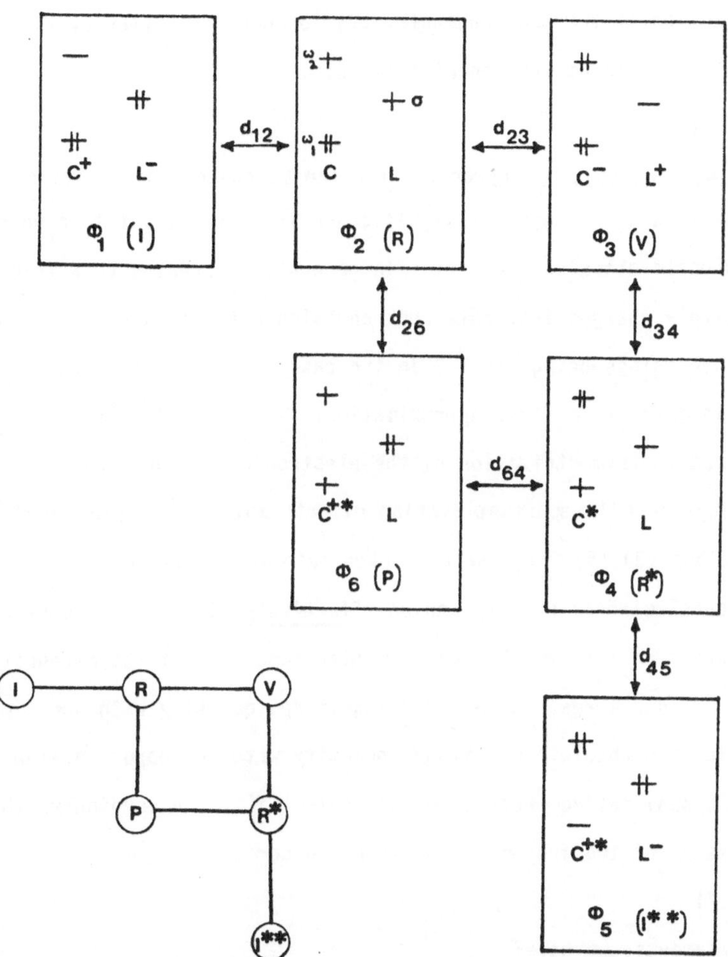

FIGURE 1: MOVB CW's, Φ_i, for the treatment of a four electron-three orbital problem. C=Core fragment and L=Ligand fragment. Fragment charges and local excitation energies are indicated by the superscripts of C and L, with one asterisk implying single local excitation and a double asterisk double local excitation. CW nomenclature is indicated in parenthesis. Double arrows indicate the monoelectronic CT interaction of the CW's through d_{ij} matrix elements. Orbital convention is spelled out in connection with the Φ_2 CW (ω_1 and ω_2 are orthogonal as they belong to the same fragment). At lower left, a schematic representation of the CT interaction of the six MOVB CW's is given with the lines connecting the circles implying d_{ij} matrix elements.

The interaction of the two fragments, L_i, brought into play by the diagonal matrix element H_{ii}, can be written as follows:

$$L_i = G_i + X_i \tag{5}$$

In eqs. (3), (4), and (5), G_i represents the (adjusted) coulomb interaction of the two fragments and its physical significance is self evident to anyone familiar with basic classical physics. In general, G_i creates an attraction between oppositely charged fragments and repulsion between identically charged fragments, other things being equal. In the case of neutral fragments, G_i can be set equal to zero, to a first approximation.

The required antisymmetrization of the electronic wavefunction brings into play the all important overlap effects which are represented by X_i in equations (3)-(5). X_i can be broken into two parts, an effectively monoelectronic part X_i' and an effectively bielectronic part X_i''. However, we have seen that the latter term plays the role of the attenuator of the former as X' is a large, in absolute magnitude, quantity with one sign and X'' is a smaller, in absolute magnitude, quantity with the opposite sign. Thus, in developing a qualitative theory, we can neglect X''. Accordingly, the interaction energy of two fragments due to H_{ii} becomes:

$$L_i = G_i + X_i' \tag{6}$$

For neutral fragments, we have:

$$L_i \simeq X_i' \tag{7}$$

Recall now that X_i' is given by the expression shown below, where $H^{o\prime}_{ab}$ represents the "monoelectronic" part of the "semiclassical" term H^{o}_{ab} over the Slater determinants X_a and X_b, EX'_{aa} represents the sum of the "monoelectronic" parts of the exchange terms over X_a and X_b, and EX'_{ab} has a meaning analogous to that of EX'_{aa}.

$$X_i' = \sum_a^k \sum_b^k \lambda_a \lambda_b H^{\circ\prime}_{ab} + \sum_a^k \lambda_a^2 EX'_{aa} + \sum_a^k \sum_b^k \lambda_a \lambda_b EX'_{ab} \tag{8}$$

We now seek to simplify EX'_{aa} and EX'_{ab}. In order to do so, we write EX'_{aa} in the form shown below, where $E'_{aa}(P^1)$ represents the sum of energy terms generated by interchange of the coordinates of an electron pair, $E'_{aa}(P^2)$ represents the sum of energy terms generated by interchange of the coordinates of two electron pairs, etc.

$$EX'_{aa} = E'_{aa}(P^1) + \underbrace{E'_{aa}(P^2) + E'_{aa}(P^3)}_{\text{ATTENUATOR TERM}} + \ldots \tag{9}$$

Because of the progressive decrease of the size of each consecutive term and the sign alternation, we can consider the algebraic sum of all terms other than $E'_{aa}(P^1)$ as the <u>attenuator</u> of the latter. Thus, in developing a qualitative theory, we can write:

$$EX'_{aa} \simeq E'_{aa}(P^1) = EX^o_{aa} \tag{10}$$

EX^o_{aa} is nothing but the principal "monoelectronic" interaction term due to the Slater determinant X_a. Similarly, we can show that

$$EX'_{ab} \simeq E'_{ab}(P^1) = EX^o_{ab} \tag{11}$$

EX^o_{ab} is the principal "monoelectronic" interaction term due to the interaction of the Slater determinants X_a and X_b. Accordingly, we can replace equation (8) with:

$$X^\circ_i = \sum_a^k \sum_b^k \lambda_a \lambda_b H^{\circ\prime}_{ab} + \sum_a^k \lambda_a^2 \, EX^\circ_{aa} + \sum_a^k \sum_b^k \lambda_a \lambda_b \, EX^\circ_{ab} \tag{12}$$

As a result, the master formula for the diagonal matrix element, H_{ii}, becomes:

$$H_{ii} = F_i + G_i + X_i^o \tag{13a}$$

or

$$H_{ii} = P_i + G_i + X_i^o \tag{13b}$$

An illustrative example is provided by the calculation of the energy of Φ shown below.

ω_2 ↑

ω_1 ↑↓ ↓ ω_3

Φ

The wavefunction can be written as:

$$\Phi = N(1\bar{1}2\bar{3} + 1\bar{1}3\bar{2}) \tag{14}$$

$$\Phi = N(X_a + X_b) \tag{15}$$

According to (13a), E is a sum of six matrix elements shown on the right hand column below obtained by truncation of five of the six rigorous matrix elements shown on the left hand column:

Rigorous Terms	
$\langle 1\bar{1}2\bar{3}\|\hat{H}\|1\bar{1}2\bar{3}\rangle$	F + G
$-\langle 1\bar{1}2\bar{3}\|\hat{H}\|2\bar{1}1\bar{3}\rangle$	$E'_{aa}\ (P_{13})$
$-\langle 1\bar{1}2\bar{3}\|\hat{H}\|1\bar{3}2\bar{1}\rangle$	$E'_{aa}\ (P_{24})$
$+\langle 1\bar{1}2\bar{3}\|\hat{H}\|1\bar{1}3\bar{2}\rangle$	$H^{o\prime}$
$-\langle 1\bar{1}2\bar{3}\|\hat{H}\|3\bar{1}1\bar{2}\rangle$	$E'_{ab}\ (P_{13})$
$-\langle 1\bar{1}2\bar{3}\|\hat{H}\|1\bar{2}3\bar{1}\rangle$	$E'_{ab}\ (P_{24})$

$$E \ \alpha\ \underbrace{[F+G+H^{o\prime} + \underbrace{[E'_{aa}(P_{12}) + E'_{aa}(P_{24})]}_{EX_{aa}^{o}} + \underbrace{[(E'_{ab}(P_{13}) + E'_{ab}(P_{24})]}_{EX_{ab}^{o}}}_{X^{o}} \tag{16}$$

Noting that since F+G is zero order, $H^{o'}$ first order, EX^{o}_{aa} second order, and EX^{o}_{ab} third order in overlap, we can neglect EX^{o}_{ab}. Henceforth, we shall be making use of formula (12) in connection with this approximation.

The X^{o} terms of representative MOVB CW's which bring into play the overlap interaction of two fragments, each containing a single orbital and zero to two electrons, are given in Table 2. The following "recipe" emerges:

1) Maximal overlap attraction, proportional to 2hs, results from the pairing of two èlectrons having opposite spins and occupying two different nonorthogonal orbitals.

2) Intermediate overlap repulsion, proportional to -2hs, results from a three electron interaction or the interaction of two electrons having the same spin and occupying two different nonorthogonal orbitals.

3) Maximal overlap repulsion, proportional to -4hs, results from a four electron interaction.

We can use this "recipe" in order to rank polyelectronic MOVB CW's according to their overlap interaction energy, X^{o}. For example, the X^{o} terms of the six MOVB CW's of Figure 1 vary in the following manner:

$$X^{o}_{1} \simeq X^{o}_{5} \simeq X^{o}_{6} > X^{o}_{2} \simeq X^{o}_{3} \simeq X^{o}_{4} \tag{17}$$

In each of Φ_2 and Φ_4, there exists one repulsive three electron interaction which is counteracted by one attractive two electron interaction, while in Φ_3 X^{o} is zero since the two electron pairs occupy orthogonal orbitals. Accordingly, Φ_2, Φ_3, and Φ_4 will have comparable X^{o} energy. In Φ_6 there exist two repulsive three-electron interactions while in each of Φ_1 and Φ_5 there exists a single repulsive four-electron interaction. Since the four electron repulsion

Table 2 Overlap Interaction Energy of Protopypical MOVB CW's.*

ω_1	ω_2	$X°$
—	↑	0
↑	↓	2 hs
↑↓	—	0
↑	↑	-2 hs
↑↓	↑	-2 hs
↑↓	↑↓	-4 hs

* $h = \langle\omega_1|\hat{O}'|\omega_2\rangle$ and $s = \langle\omega_1|\omega_2\rangle$, with $\hat{O}'$ being the monoelectronic part of the Hamiltonian operator.

is twice as large as the three electron repulsion, Φ_1, Φ_5 and Φ_6 will also have comparable X^o energy and this will be higher than the X^o energy of the other three CW's.

Turning our attention to MOVB interaction matrix elements, H_{ij}'s, we note that the classification of CW's according to excitation type provides the basis for a classification of interaction matrix elements according to the type of CW's which they connect. Specifically, we may distinguish the following five types of H_{ij}'s:

a) The two CW's, i and j, differ by one occupied spin orbital with m and n being two orbitals of different fragments differentially occupied in the i and j CW's. The interaction matrix element, symbolized by d_{ij}, is a Charge Transfer (CT) matrix element.

b) The two CW's differ by one occupied spin orbital with m and n being two orbitals of one and the same fragment differentially occupied in the i and j CW's. The interaction matrix element, symbolized by p_{ij} is a polarization matrix element which brings about the interaction of two spin orbitals m and n which were originally orthogonal in the isolated fragment but which can now mix under the influence of a second fragment.

c) The two CW's differ by two occupied spin orbitals with m,n and p,q being orbital pairs of different fragments differentially occupied in the i and j CW's. The interaction matrix element, symbolized by D_{ij}, is a bielectronic CT matrix element.

d) The two CW's differ by two occupied spin orbitals, with m, $\bar{m}$ and n,$\bar{n}$ being orbital pairs of one and the same fragment differentially occupied in the i and j CW's. The interaction matrix element, symbolized by P_{ij}, is a bielectronic polarization matrix element which brings about the interaction of four originally orthogonal spin orbitals of one and the same fragment.

e) The two CW's differ by two occupied spin orbitals and their interaction represents two coupled single excitations of the two fragments. The interaction matrix element, symbolized by W_{ij}, is a <u>bielectronic correlation matrix element.</u> Mixed CT-polarization matrix elements are not considered.

A pictorial depiction of the five types of interaction matrix elements is given in Figure 2. The CT matrix elements d_{ij} and D_{ij} are responsible for one and two electron transfers, respectively, from one fragment to the other. The polarization matrix elements are responsible for local mono- (p_{ij}) and di-excitation (P_{ij}). The very nature of these matrix elements defines two entirely different problems:

a) Cases where interfragmental spatial overlap is large, e.g., ground state equilibrium geometries. Here, bonding is primarily controlled by the d_{ij} and secondarily by the related D_{ij} matrix elements, since both of them are functions of interfragmental AO overlap.

b) Cases where interfragmental spatial overlap is nearly zero. Here, any bonding due to CI is primarily due to the p_{ij} and W_{ij} matrix elements which depend on the (long range) coulomb field of the two fragments.

In this work, we shall be interested in applications of MOVB theory to ground state equilibrium geometries of composite systems made up of strongly interacting fragments, e.g., organic molecules. Such a treatment has never been attempted before. By contrast, MOVB theory has already been applied to the problem of long range molecular interaction.[21] Thus, our next goal is to simplify the interaction matrix elements so that their functional dependence is always explicitly evident, paying particular attention to d_{ij} CT matrix elements.

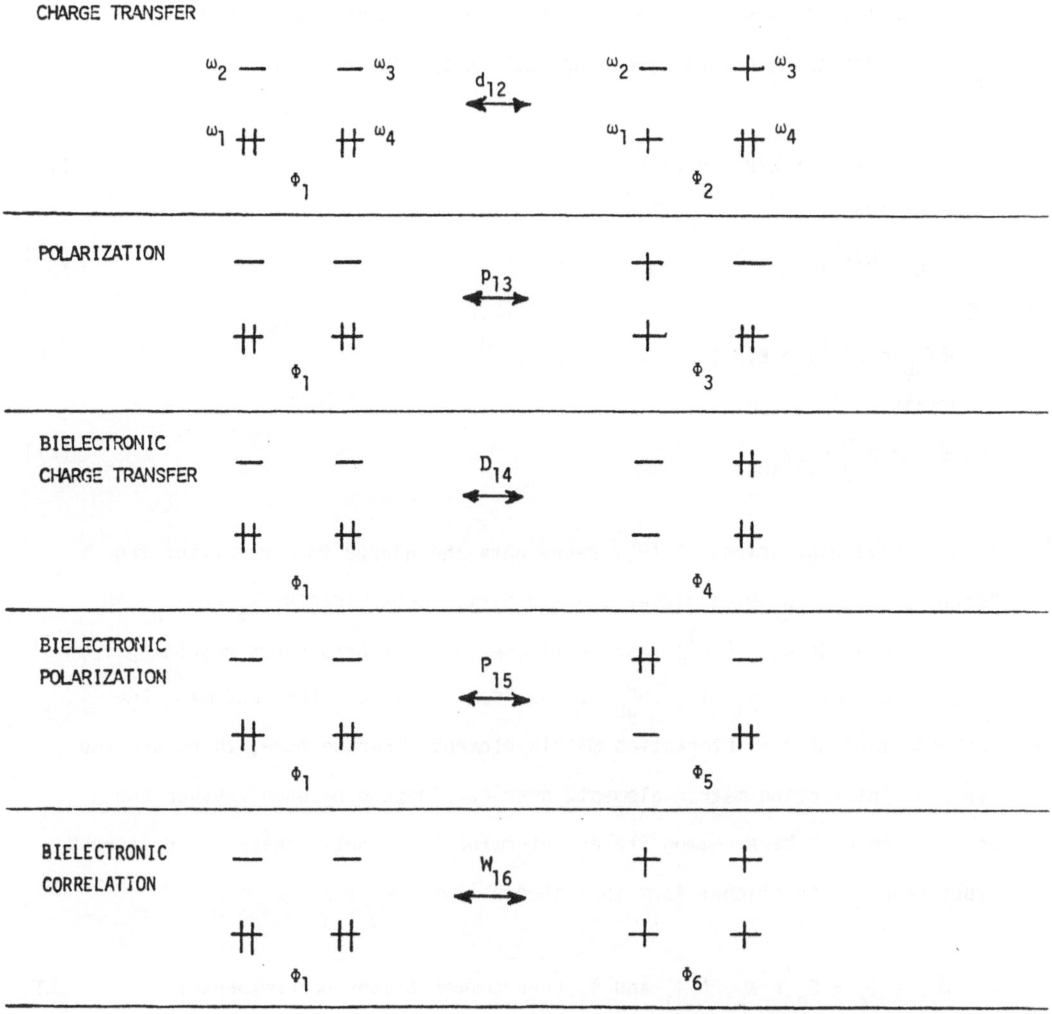

CHARGE TRANSFER

ω_2 — — ω_3 d_{12} ω_2 — + ω_3

ω_1 ⧺ ⧺ ω_4 ⟷ ω_1 + ⧺ ω_4

Φ_1 Φ_2

POLARIZATION

p_{13} Φ_1 Φ_3

BIELECTRONIC CHARGE TRANSFER

D_{14} Φ_1 Φ_4

BIELECTRONIC POLARIZATION

P_{15} Φ_1 Φ_5

BIELECTRONIC CORRELATION

W_{16} Φ_1 Φ_6

FIGURE 2: The prototypical interaction matrix elements of MOVB theory.

As we have discussed in Part I, the interaction matrix element, H_{ab}, over Slater determinants can be cast in the following forms:

$$H_{ab} = E(P^o) + E(P^1) + E(P^2) + \ldots \qquad (18)$$

If we define

$$H_{ab}^{o} = E(P^o) \qquad (19)$$

and

$$EX_{ab} = E(P^1) + E(P^2) + \ldots \qquad (20)$$

we obtain

$$H_{ab} = H_{ab}^{o} + EX_{ab} \qquad (21)$$

In the above expressions, E (P^o) represents the energy term resulting from a "zero" permutation which places the two Slater determinants, X_a and X_b, in maximum coincidence. $E(P^1)$ represents the sum of energy terms resulting from "single" permutations, etc. H_{ab}^{o} is the "semiclassical" term and EX_{ab} the exchange term of the interaction matrix element. Furthermore, there are two types of interaction matrix elements over CW's depending upon whether the interacting CW's have common Slater determinants or not. These two different types have the functional form indicated below:

$$H_{ij} = F_c + G_c + X_{ij} \quad (\Phi_i \text{ and } \Phi_j \text{ have common Slater Determinants}) \qquad (22)$$

$$H_{ij} = X_{ij} \quad (\Phi_i \text{ and } \Phi_j \text{ do not have common Slater Determinants}) \qquad (23)$$

By working in the same manner as before, we produce the following master formulae, respectively:

$$H_{ij} = F_c + G_c + X_{ij}^{o} \quad (\Phi_i \text{ and } \Phi_j \text{ have common } X_a\text{'s}) \qquad (24)$$

$$H_{ij} = X_{ij}^{o} \quad (\Phi_i \text{ and } \Phi_j \text{ do not have common } X_a\text{'s}) \qquad (25)$$

Interaction matrix elements of the first type connect spin degenerate CW's and they can be neglected, for qualitative purposes, if we take advantage of spin degeneracy and construct linearly independent CW's which differ substantially in energy and which, thus, can interact only to a small extent. For example, consider the two independent four open shell electron wavefunctions Φ_1 and Φ_2 shown below[22]:

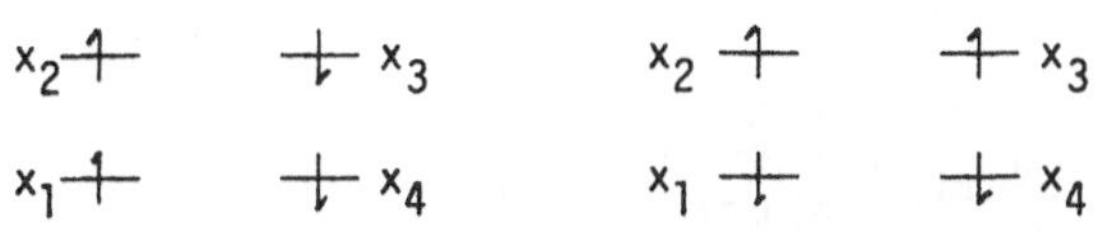

$$x_2 \uparrow \quad \downarrow x_3 \qquad x_2 \uparrow \quad \uparrow x_3$$

$$x_1 \uparrow \quad \downarrow x_4 \qquad x_1 \downarrow \quad \downarrow x_4$$

$$\Phi_1 \propto X_a - X_b - X_c + X_d \qquad \Phi_2 \propto X_a - X_e - X_f + X_d$$

Now, in MOVB theory with core-ligand dissection, intrafragmental MO overlap is zero.[23] Furthermore, in most applications, only two pairs of MO's overlap strongly in an interfragmental sense (x_1-x_4 and x_2-x_3, or, x_1-x_3 and x_2-x_4), with the other two pairs overlapping weakly or not at all. In such cases, one can easily define Φ_1 and Φ_2 so that the first generates interfragmental bonds and the latter interfragmental antibonds due to spin pairing. Because of their large energy separation, H_{12} can then be neglected. With this in mind, we will now proceed to develop the theory by focusing attention on the qualitatively important interaction matrix elements of the second type, i.e., equation (25).

Equation (25) represents a reasonable approximation of interaction matrix elements comprised of monoelectronic <u>and</u> bielectronic terms, i.e., d_{ij}, D_{ij}, and p_{ij} interaction matrix elements. On the other hand, it cannot reveal the functional dependence of interaction matrix elements containing only bielectronic parts, such as P_{ij} and W_{ij}, because these are reduced to zero according to equation (25). Recall that the forms of (22) and (23) were dictated at the VB theoretical level by the fact that interaction matrix elements of both types are important at the VB theoretical level and the desire to make consistent approximations throughout. At the MOVB theoretical level, we have the luxury of being concerned with only one type of interaction matrix element, that between CW's having no common Slater determinants. Thus, we can start anew from the formal expression of H_{ij} shown below:

$$H_{ij} = \sum_{a}^{k} \sum_{b=\ell}^{z} \lambda_a \lambda_b H^{\circ}{}_{ab} + \sum_{a}^{k} \sum_{b=\ell}^{z} \lambda_a \lambda_b EX_{ab} \tag{26}$$

We can set EX_{ab} equal to zero, since the component terms are small (third order in overlap terms). However, instead of neglecting the "bielectronic" terms of H^{o}_{ab}, we approximate H^{o}_{ab} by truncating its expansion to the largest term. In this way, we are assured of the correct functional dependence of H_{ij}.

$$H_{ij} = \sum_{a}^{k} \sum_{b=\ell}^{z} \lambda_a \lambda_b Q^{\circ}{}_{ab} \tag{27}$$

where Q^{o}_{ab} is the truncated form of the "semiclassical" matrix element H°_{ab}. This expression will be used throughout this series of papers, unless otherwise stated. Application of equation (27) to the representative interaction matrix elements pictorially depicted in Figure 2 produces the results given below:

$$d_{12} \propto \langle \omega_1 | \hat{O}' | \omega_3 \rangle = h_{13} \tag{28}$$

$$P_{13} \propto \langle \omega_1 | V_B | \omega_2 \rangle \tag{29}$$

$$D_{14} \propto \langle \omega_1 | \hat{O}' | \omega_3 \rangle \langle \omega_1 | \omega_3 \rangle \tag{30}$$

$$P_{15} \propto \langle \omega_1 \omega_2 | \omega_1 \omega_2 \rangle \tag{31}$$

$$W_{16} \propto \langle \omega_1 \omega_2 | \omega_3 \omega_4 \rangle \tag{32}$$

In the above expressions $\hat{O}'$ is the monoelectronic part of the Hamiltonian, V_B is the core operator of fragment B acting on the electron distribution of fragment A, and the rest of the terms have their usual significance. Special attention ought to be paid to the CT matrix element, d_{ij}, which can be approximated as follows:

$$d_{ij} = h_{mn} \tag{33}$$

with

$$h_{mn} = \langle m|\hat{O}'|n\rangle \tag{34}$$

where m and n are the two MO's which differ in occupancy by one electron in Φ_i and Φ_j. Each h_{mn} can be expanded into a sum of integrals over AO's, denoted by the subscripts t and u:

$$h_{tt} = \langle t|\hat{O}'|t\rangle \tag{35}$$

$$\beta_{tu} = h_{tu} = \langle t|\hat{O}'|u\rangle \tag{36}$$

The AO resonance integral, β_{tu}, can be approximated by any one of several proposed expressions, such as the Wolfsberg-Helmholz formula:[24,25]

$$\beta_{tu} = K\,(h_{tt} + h_{uu})\,s_{tu} \tag{37}$$

In the above expression, K is an energy constant and s_{tu} the AO overlap integral.

Parenthetically, we note that the form of the Wolfsberg-Helmholz AO interaction matrix element shown above defines two extreme categories of atoms:

a) Weakly bonding atoms characterized by low electronegativity (i.e., small h_{tt}) and poor spatial orbital extension (i.e., small s_{tu}). In general, weakly bonding atoms are to be found towards the left of a period and towards the bottom of a column of the Periodic Table, e.g., Li and I.

b) Strongly bonding atoms, characterized by high electronegativity (i.e., large negative h_{tt}) and good spatial orbital extension (i.e., large s_{tu}). In general, strongly bonding atoms are to be found towards the right of a period and towards the top of a column of the Periodic Table, e.g., O and F.

The relative sizes of matrix elements H_{ab} can be immediately deduced, in a qualitative sense. On the other hand, the sign of each individual matrix element over VB-type CW's cannot be predicted in any simple manner, at least when the VB-type CW's are polyelectronic functions describing open shell electrons. We have already seen that in VB theory stereoselection is expressed via the signs as well as magnitudes of matrix elements, which, in turn, depend on the signs of the AO overlap integrals. We say then that <u>VB theory is a sign dependent theory</u>. The "sign problem" constitutes one of the main obstacles blocking our way towards the development of a qualitative VB theory of real systems. The great advantage of MOVB theory is that it eliminates this problem and paves the way towards a detailed understanding of chemical bonding in polyatomic systems. Let us now examine how the sign independence of MOVB theory arises.

There exist four distinct types of VB CW matrix elements:

a) Diagonal matrix elements, H_{ii}. The form of these matrix elements is:

$$H_{ii} \alpha P_i + G_i + f(s^{even}) + f(s^{odd}) \quad (38)$$

where P_i is the promotional energy, G_i is the "classical" interaction energy, $f(s^{even})$ is the overlap energy contribution which is dependent upon even powers of AO overlap integrals, s_{tu}'s, and $f(s^{odd})$ is the overlap energy contribution which is dependent upon odd powers of the AO overlap integrals, s_{tu}'s. Now, when the AO's overlap in a cyclic manner, the magnitude of H_{ii} depends on the signs of s_{tu}'s.[26] If there is no cyclic AO overlap, H_{ii} is independent of the signs of s_{tu}'s.

b) Off-diagonal matrix elements over linearly independent CW's which describe identical orbital occupancy. The form of these matrix elements is the same as above. Accordingly, we arrive at the same conclusions as in (a).

c) Off-diagonal matrix elements over CW's which differ in orbital occupancy by two spin orbitals, $H_{ij}(2)$. The approximate form of these matrix elements is:

$$H_{ij}(2) \alpha f(s^{even}) + f(s^{odd}) \quad (39)$$

d) Off-diagonal matrix elements over CW's which differ in orbital occupancy by one spin orbital, $H_{ij}(1)$. The approximate form of these matrix elements is:

$$H_{ij}(1) \alpha f(s^{odd}) \quad (40)$$

Now, when the AO's overlap in a cyclic manner, the magnitude <u>and</u> sign of H_{ij} are dependent upon the signs of the s_{tu}'s. On the other hand, when the AO's overlap in a noncyclic manner, there is magnitude independence but sign dependence upon the signs of s_{tu}'s.

It is then clear that stereoselection is expressed via signs and magnitudes of matrix elements when AO's overlap in a cyclic manner. On the other hand, we saw that noncyclic AO overlap may also lead to sign dependence of the matrix elements which, at first sight, seems to imply that stereoselection can arise even when AO's overlap in a noncyclic manner. However, this is not the case as there is another consequence of AO overlap which has already been mentioned: Cyclic AO overlap gives rise to a cyclic interaction of CW's such that the product of CW interaction matrix elements defining the cyclic CW interaction, Π, is an odd function of s_{tu}'s. On the other hand, noncyclic AO overlap may give rise to a cyclic overlap of CW's but now the product of the CW interaction matrix elements defining the cyclic CW interaction is an even function of each s_{tu}. For example, the noncyclic system shown below containing two electrons can be approximately described by six low energy open shell CW's which interact in a <u>cyclic</u> manner. However, the product of the H_{ij}'s is an even function of β_{tu}'s and s_{tu}'s and, thus, the system is truly nonaromatic as expected (see Figure 6 of part I).

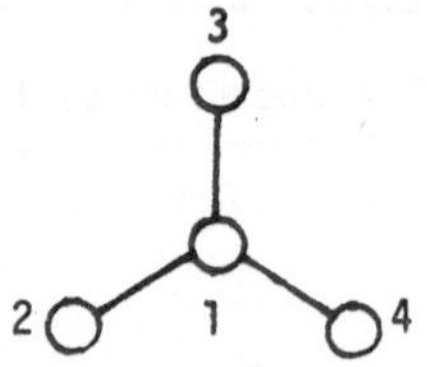

3

2 1 4

2e

In general, cyclic orbital overlap implies:

a) Sign and magnitude dependence of matrix elements on the signs of orbital overlap integrals. In this case, we have:

$$\Pi_c \;\alpha \pm H_{ij} \cdot H_{jk} \cdot H_{kl} \cdots H_{zi} \tag{41a}$$

or

$$\Pi_c \;\alpha \pm s_{ab}^{odd} \cdot s_{cd}^{odd} \cdots s_{kl}^{even} \cdot s_{mn}^{even} \cdots \tag{41b}$$

with $\Pi>0$ producing a set of Hückel and $\Pi<0$ producing a set of Möbius eigenstates.

On the other hand, noncyclic orbital overlap implies:

a) Magnitude independence of matrix elements from the signs of orbital overlap integrals.

b) $\Pi_n = 0$ (42)

or

$$\Pi_n \;\alpha \pm s_{ab}^{even} \cdot s_{cd}^{even} \cdots \tag{43}$$

Let us now consider what (41) - (43) imply with regards to the role that H_{ij} signs play in VB theory. In the case of cyclic orbital overlap, a positive s_{ab} can represent geometry A^+ and a negative s_{ab} can represent geometry A^- because Π_c is a function of s_{ab}^{odd}. Furthermore, a determination of the sign of each H_{ij} is mandatory because these determine the sign preceding Π_c and, ultimately, whether A^+ involves Hückel and A^- Möbius CW interaction, or vice versa. By contrast, <u>if the original set of AO's is transformed into a new set of orbitals which cannot overlap in a cyclic manner</u>, then geometries A^+ and A^- can no longer be represented by two different sign allocations to s_{ab} since Π_n is either zero or

s_{ab} sign invariant as it is a function of s_{ab}^{even}. It follows that determination of the signs of the H_{ij}'s is unnecessary and that the two geometries are now distinguishable on the basis of some criterion other than overlap integral signs. This suggests that we can bypass the necessity of figuring out H_{ij} signs in qualitative applications if we develop the VB theory of chemical bonding starting with a basis set of noncyclically overlapping orbitals. MOVB theory is indeed founded on such a principle since the fragment MO's cannot interact in a cyclic manner, being orthogonal in a intrafragmental sense. For example, the VB treatment of the pi system of cyclobutadiene necessitates the proper determination of the signs of matrix elements because the four 2p AO's overlap in a cyclic manner. By contrast, the MOVB treatment of the same system over a set of two π and two π^* MO's spanning two fragment pi ethylenes does not hinge on the signs of matrix elements because π_1 and π_1^* as well as π_2 and π_2^* are orthogonal in an intrafragmental sense. This is schematically illustrated below. We shall see that, in MOVB theory, stereoselection arises not as a result of different assignments of the signs of the MO overlap integrals, but, rather, as a result of different excitation patterns dictated by the symmetry of the fragment MO's, which, in turn, depends on the assignment of the signs of the AO overlap integrals.

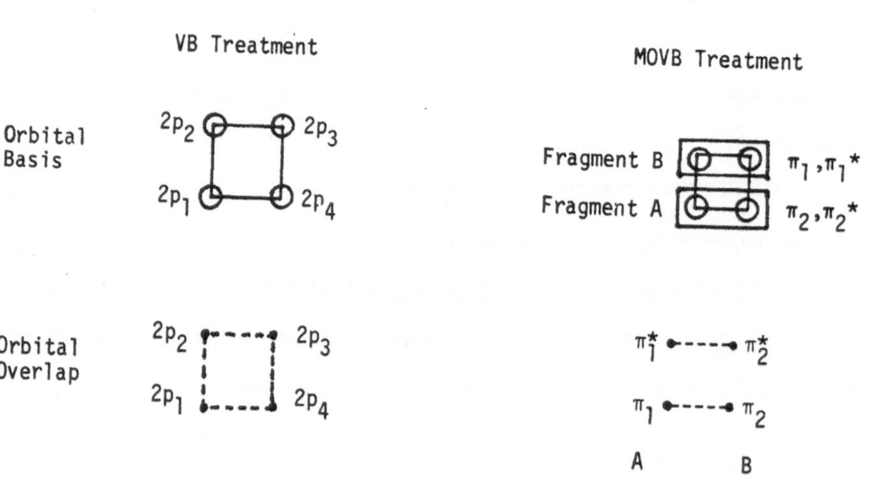

In closing, we note that when cyclic AO overlap is rendered untenable by symmetry constraints or zero overlap,the dependence of the total energy expressions on the signs of matrix elements ceases to exist and one may implement VB theory in connection with a Core-Ligand dissection which ensures intrafragmental orbital orthogonality. In such a case, the conceptual frameworks of VB and MOVB theory become identical. A typical example is the linear interaction of three nondegenerate AO's x_1, x_2, x_3 containing a given number of electrons. Since x_1 and x_3 do not overlap, the choice of fragments indicated below makes possible the application of VB theory with Core-Ligand dissection because x_1 and x_3 are orthogonal due to zero spatial overlap.

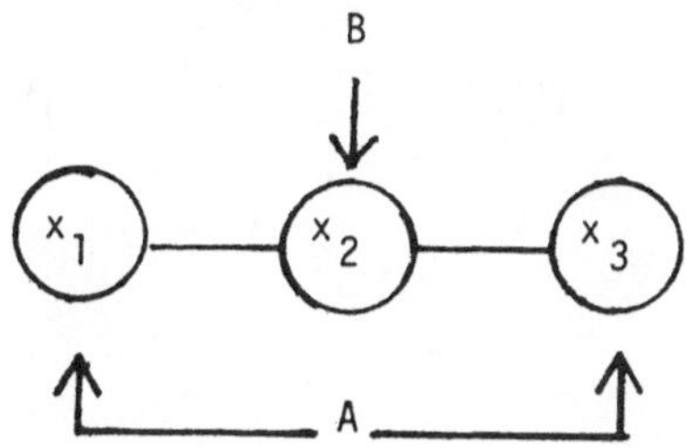

B. The MOVB Approach to Ground State Stereochemistry

Consider a four electron composite system CL, made up of a core, C, and a set of ligands, L, with two orbitals spanning the core and one orbital spanning the ligands. We ask the question: What is the stereochemistry which permits optimum ground state bonding of C and L? Stated in different language, what is the point group symmetry of the CL system which dictates the point group symmetry of the three orbitals, which, in turn, allows the three orbitals to best accomodate the four electrons? We can imagine three different geometries of CL which impose the orbital symmetry relationships depicted below:

C	L	C	L	C	L
$\omega_2(\Gamma_a)$ ___		$\omega_2(\Gamma_a)$ ___		$\omega_2(\Gamma_b)$ ___	
	___ $\sigma(\Gamma_a)$		___ $\sigma(\Gamma_a)$		___ $\sigma(\Gamma_a)$
$\omega_1(\Gamma_b)$ ___		$\omega_1(\Gamma_a)$ ___		$\omega_1(\Gamma_a)$ ___	
I		II		III	

Γ_a denotes the point group irreducible representation according to which an orbital transforms. In simpler language, ω_1 may have A and ω_2 and σ, S symmetry with regard to a symmetry element in geometry I, etc. With the aid of Figure 3, we can now understand the following:

a) In geometry I, a single bond connecting C and L is formed by coupling an electron in ω_2 and another in σ while the remaining electron pair is allowed to occupy the low energy ω_1 of C and is unable to be delocalized in any of the other two orbitals by symmetry.

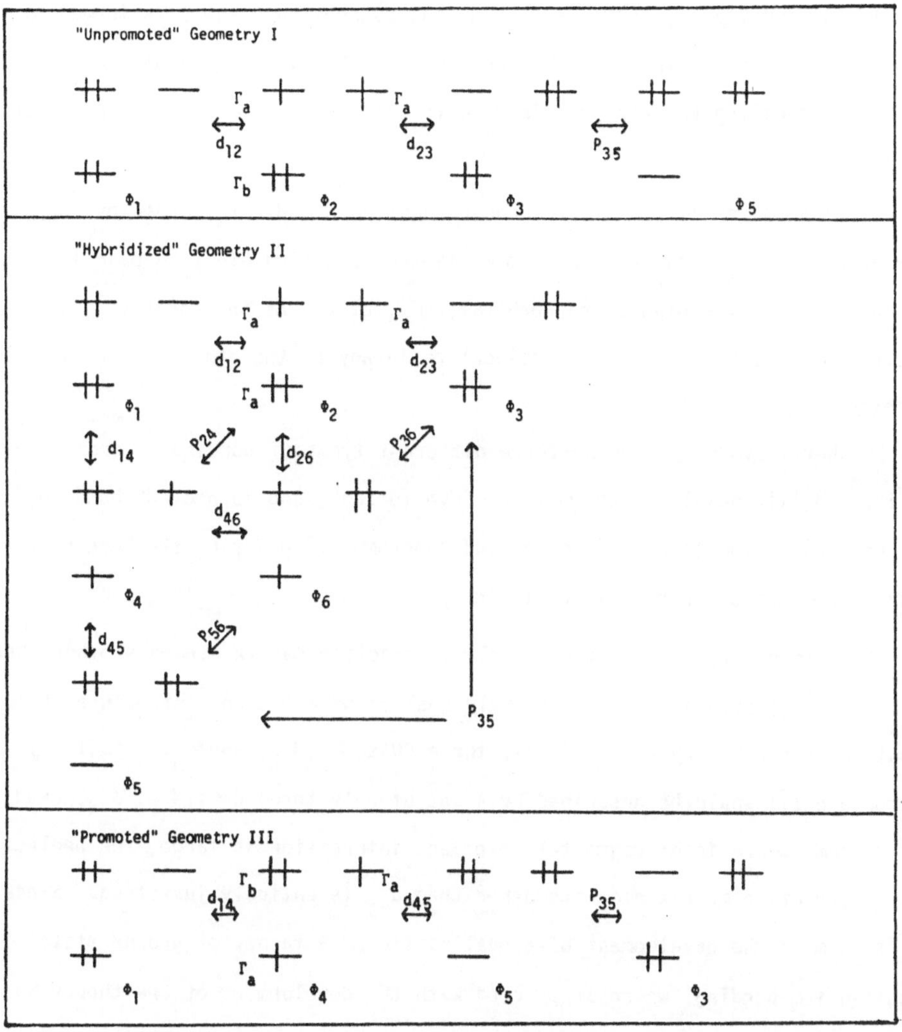

FIGURE 3: Chemical bonding in C̈-L under three different orbital symmetry constraints. Note how the symmetry constraints produce three different manifolds of interacting CW's in each case.

b) In geometry II, a single <u>hybrid</u> bond connecting C and L is formed by coupling one electron in ω_1 and another in σ and keeping ω_2 doubly occupied as well as by coupling one electron in ω_2 with another in σ and keeping ω_1 doubly occupied.

c) In geometry III, a single bond connecting C and L is formed by coupling an electron in ω_1 with another in σ while the remaining electron pair is forced to occupy the high energy orbital, ω_2, of C. As in geometry I, the electron pair is still unable to delocalize in any of the other two orbitals by symmetry.

In short, symmetry imposes three different types of bonding in geometries I, II, and III, namely, "unpromoted", "hybridized", and "promoted" bonding, respectively. The terms "promoted" and "unpromoted" describe the lone electron pair accommodation in geometries I and III.

At this point, we note that if all interaction matrix elements other than d_{ij} were neglected as being relatively small, the electronic structure of I would be described by a set of only three CW's, Φ_1, Φ_2, and Φ_3. Similarly, structure III would be described by a set of only three CW's, Φ_3, Φ_4, and Φ_5. In systems where interfragmental overlap interaction is large, the neglect of all interaction matrix elements other than d_{ij} is entirely justified. Since our first aim is the development of a qualitative MOVB theory of ground state equilibrium bonding, we shall proceed with the development of the theory by assuming that CW interaction can be effected only through the CT interaction matrix element, d_{ij}. When interfragmental overlap interaction is small either as a result of weak bonding potential of the constituent atoms [i.e., small negative h_{tt} in equation (37)], or, small interfragmental <u>spatial</u> overlap [i.e., small s_{tu} in equation (37)] all matrix elements must be considered. Thus, in developing the theory of strong bonding we should bear in mind the necessary modifications and additional considerations necessitated in the case of weak bonding.

With the above stated assumptions in mind, let us now recall the treatment of the four electron-three orbital problem at the VB level. There, we saw that we could define three elementary VB structures, each made up of one elementary bond and one elementary lone pair, which could be coupled in an intrinsic, direct and indirect manner in all three possible stereochemical arrangements simulated by assigning a negative sign, a zero, or a positive sign to an AO overlap integral. The relative energies of the three stereochemical arrangements were interpreted to be reflections of relative coupling efficiency. This simple VB model of chemical stereoselection is replaced at the MOVB theoretical level by a related model which, in fact, is more revealing, in a chemical sense. Thus, geometry I is now compatible with a single MOVB elementary structure, M, geometry III is compatible with a second distinct elementary structure, N, and geometry II is compatible with two coupled elementary structures M and N plus the additional MOVB CW P. In geometry II, the coupling of M and N is intrinsic via the sharing of V as well as indirect via interaction of each M and N with P. These considerations are illustrated schematically below, where the lines connecting the circles represent CT matrix elements and the circles themselves represent the denoted CW's.

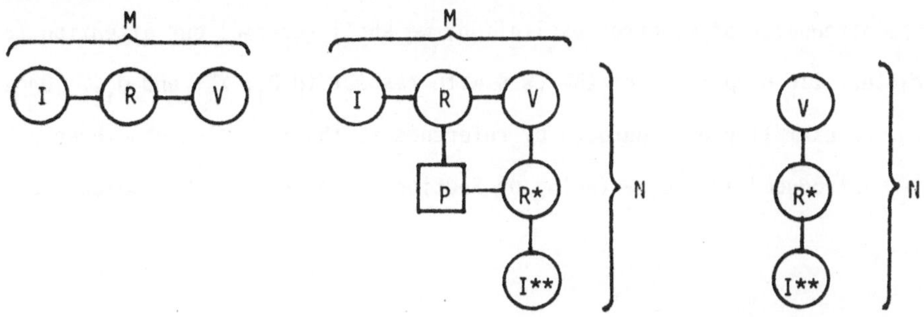

It is evident that excitation energy considerations favor geometries I and II over III and that the relative energy of I and II will depend on the coupling efficiency of II. Thus, we have already taken a glimpse of the general MOVB stereoselection model which we shall begin to develop in the next sections. Before we do so, however, let us review briefly the critical energetic characteristics of the six CW's needed for the treatment of the four electron-three orbital system so that we begin to develop an intuition as to what the roles of different types of CW's will be in the stereochemical problems we shall tackle at the end of Part I.

Let us recall equations (13) and (27) which constitute the basis of qualitative MOVB theory. According to these equations, the critical factors which need to be considered are the following:

1) The excitation energy of a CW, i.e., P_i in equation (13b).

2) The "classical" coulomb interaction of the two fragments, i.e., G_i in equations(13).

3) The overlap interaction of the two fragments, i.e., X_i^o in equations(13).

4) The interaction of a given CW with others, i.e., H_{ij} in equation (27).
For problems of ground molecular structure, one can restrict attention to interaction via d_{ij} which is proportional to the interaction of the two MO's which define the "origin" and "terminus" of the electron transfer.
It is evident that each CW has characteristic properites with respect to each of the above factors. Since " classical" coulomb interaction can be viewed as the attenuator of electron excitation, we shall restrict our attention to the characteristic properties of the CW's with respect to P_i, X_i^o, and d_{ij}. Once again, we exemplify our approach by reference to the four electron-three orbital system introduced in the beginning of Section A and the six CW's which are

required for a complete electronic description and which are shown in Figure 1.

The characteristic properties of the CW's insofar as excitation energy is concerned are quite obvious. If we define the Φ_2 CW as the reference CW, then all other CW's represent CT excitation, local excitation,or, combinations thereof. One point that needs attention is the way in which P_i can be computed or estimated. Specifically, this quantity is determined by evaluating the energy of each CW in the absence of interfragmental interaction, i.e., F_i in equation (3), and writing their relative energies in a way that zero energy is assigned to the lowest energy CW. Now, F_i is a sum of mono- and bi-electronic terms. If the latter terms are neglected F_i is the simple sum of one-electron orbital energies. In general, excitation energies assigned on the basis of the one-electron orbital energies of the fragments are qualitatively reliable when the fragment orbitals are well separated in energy. Otherwise, bielectronic terms may reverse the order predicted on the basis of exclusive consideration of monoelectronic terms. In general, severe interelectronic repulsion accompanies the occupation of MO's which have common AO's and this must be taken into consideration when ranking CW's according to P_i of equation (13b). Fortunately, a precise knowledge of P_i is not required in the vast majority of qualitative applications of the theory to ground stereochemical problems.

The characteristic properties of the six CW's insofar as overlap attraction or repulsion of C and L is concerned are the following:

a) In the presence of symmetry constraints, e.g., geometries I and III, the closed shell CW's I,I**, and V generate neither overlap attraction nor overlap repulsion. By contrast, the open shell CW's R and R* generate overlap attraction. The P CW is "screened out" as it cannot interact with any other CW.

b) In the absence of symmetry constraints, e.g., geometry II, the closed shell CW V as well as the open shell CW's R and R* generate either zero or small overlap interaction for two entirely different reasons. In V, two electron pairs are placed in orthogonal orbitals so that there is no overlap interaction between them, while in R and R* an attractive two electron overlap interaction is counteracted by a repulsive three electron overlap interaction. The situation changes drastically in the case of the closed shell CW's I and I** as well as the open shell CW P. These CW's generate strong overlap repulsion again for entirely different reasons. In I and I**, two electron pairs are placed in overlapping orbitals and this generates a four-electron overlap repulsive interaction, while in P the four electrons are distributed among the three orbitals in a fashion which generates two three-electron overlap repulsive interactions. Thus, overlap repulsion is comparable in I, I**, and P.

Finally, the characteristic properties of the six CW's insofar as CT is concerned are the following:

a) In the presence of symmetry constraints, e.g., geometries I and III, hybridization cannot occur and two types of C-L bonds are defined by two sets of CW's. The first, is defined by the I,R, and V set (geometry I) and the second by the I**, R*, and V set (geometry III).

b) In the absence of symmetry constraints, e.g., geometry II, hybridization occurs via the interactions of two sets of CW's, the I,R,V set and the I**, R*, V set, and the additional involvement of P. Accordingly, we can distinguish the following four types of CW's:

1) A CW, V, which is common to the two sets of CW's and can be regarded as the agent of intrinsic coupling of two elementary structures, one defined by the I, R, and V set and the other by the I**, R*, and V set. We can say that V acts

as a hybridization valve for reasons that will become apparent later on. Hence, the designation V. Note that the characteristic property of a V-type CW is the "piling up" of electrons on one or the other fragment, a pattern which ensures zero interfragmental overlap repulsion.

2) A CW, P, which can bring about the indirect coupling of the two sets defining the two elementary structures by virtue of interacting directly with R, belonging to one set, and R*, belonging to the second set of CW's. Again, we can say that P acts as a hybridization valve. In addition, it introduces local excitation, or, polarization, of the fragments. Hence, the designation P. Note that the P CW enters the total wavefunction via two different types of matrix elements, namely d_{ij} and p_{ij}. Mixing of the former type can be called CT induced polarization while mixing of the latter type can be called coulomb field induced polarization. Only the former type is contained in ordinary Extended Hückel MO computations.

3) Two CW's R and R*, which belong to different sets, with the first being the "unpromoted" form of the "promoted" second one, and with both having radicaloid character. Hence, the designations, R and R*, respectively. R and R* can interact indirectly via V or via P.

4) Two CW's, I and I**, which belong to different sets, with the first being the "unpromoted" form of the "promoted" second one, with both having closed shell character. I and I** can interact with each other indirectly only via R, P and R* or via R, V and R*. As we shall see, these CW's act as hybridization insulators. Hence, the designation I and I**. Note that the characteristic property of an I-type CW is the equitable distribution of electrons to orbitals of both fragments, a pattern which ensures maximal interfragmental overlap repulsion.

MOVB theory allows the immediate recognition of the central role that orbital symmetry plays in any chemical problem and we have already identified three different prototypical bonding modes by exploiting this advantage of MOVB theory. We must now attempt to develop a representation of bonding of model systems which can be easily adapted to polyelectronic systems and which can be routinely applied to problems of interest. This key construct is the pictorial _bond diagram_. The concept of the _ground state bond diagram_ is developed in the following section.

C. Ground State Bond Diagrams

As the simplest illustrative case of our ideas, let us consider the union of two fragments C· and L· to form the two electron-two orbital system C-L. The ground state of this system, Ψ, is described by an optimal linear combination of the three CW's as shown below:

$$\Psi \alpha\, C\cdot \; \cdot L + \lambda_1[C\overset{-}{:}\; L^+] + \lambda_2[C^+\; L\overset{-}{:}] \qquad (44)$$

Instead of writing down the above equation, we can use the symbolic notation shown below:

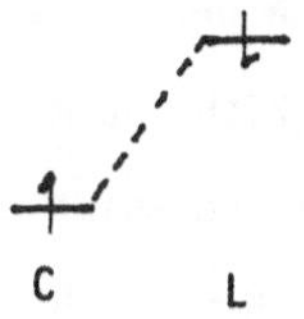

C L

The above diagram is termed a ground state bond diagram. It represents the optimum ground state wavefunction which is a linear combination of all CW's which have the proper symmetry and which can be generated by permuting electrons amongst fragment orbitals in a way made pictorially self evident by the ground state bond diagram itself. Excited state bond diagrams can be constructed in a similar fashion. However, in this paper, we shall be concerned exclusively with the development of ground state MOVB theory and we shall defer the presentation of the MOVB theory of excited states to a following paper. Consequently, we shall henceforth refer to ground state bond diagrams simply as bond diagrams.

Let us now discuss in some detail how the above bond diagram was constructed and what are some of its key implications. First of all, it is evident that the electron arrangement in the two orbitals as presented in the bond diagram reflects the CW which allows for the maximum number of electron pairings consistent with formation of the maximum number of two electron bonds. This "parent" CW is termed the reference "Perfect Pairing" (PP) CW and it is denoted by R. Second, the bond diagram is actually constructed beginning from the R CW and adding dotted lines connecting the core and ligand orbitals of the same symmetry in order to denote all additional CW's which can be generated by permuting the electrons among these orbitals. Thus, a bond diagram makes explicit the nature of the R CW, which, in most practical applications, is the lowest energy CW of the entire set and which plays a role analogous to that played by the Heitler-London (HL) CW's in VB theory. In addition, it represents schematically all CW's connected by monoelectronic d_{ij} matrix elements which contribute to the total wavefunction of the ground state.

At this point, we introduce for the first time symbolic notation to be used in connection with the bond diagrams. Thus, in the case of a two electron-two orbital bond, we shall distinguish between an N bond and an N' bond, where N indicates a fully formed and N' a weakened or broken bond. At the limit of a completely broken bond, N' is represented by the lowest energy CW of the three possible CW's.

Next, we consider the union of $\ddot{\mathrm{C}}\cdot$ and $\mathrm{L}\cdot$ to form the four electron-three orbital system $\ddot{\mathrm{C}}$—L In this case, the two-electron bond is being "observed" by an electron pair and an approximate description of this four electron-three orbital system requires six CW's which interact in the manner shown in Figure 1. Once we have more than two orbitals to deal with, our wavefunction becomes subject to symmetry control and we distinguish three possible situations:

a) Orbitals ω_2 and σ are of the same symmetry species.

b) All three orbitals are of the same symmetry species.

c) Orbitals ω_1 and σ are of the same symmetry species.

The three different bond diagrams which correspond to the three cases described above are depicted in Figure 4. The designations D, H, and U are conceived so that they are evocative of the following three different physical situations:

a) In D, the electron pair is "allowed" to relax to the lower energy orbital of C and the electron pair bond itself is formed by the overlap of the higher energy orbital ω_2 and orbital σ (D implies "down").

b) In U, the electron pair is "forced" to be confined in the higher energy orbital ω_2 of C and the electron pair bond itself is formed by the overlap of the lower energy orbital ω_1 and orbital σ (U implies "up").

c) H represents a case wherein the U and D bonding modes are "intertwined", i.e., the H system is a <u>hybrid</u> of the D and U systems.

Next, we consider the union of $\overset{o}{C}\cdot$ and L$\cdot$, where the empty circle on C denotes the presence of an unoccupied orbital, to form the two electron-three orbital system $\overset{o}{C}$—L The requisite CW's are shown in Figure 5 and the three different bond diagrams are also displayed in Figure 5. The designations D, H, and U are entirely analogous to the ones we have used in the case of the four electron-three orbital systems.

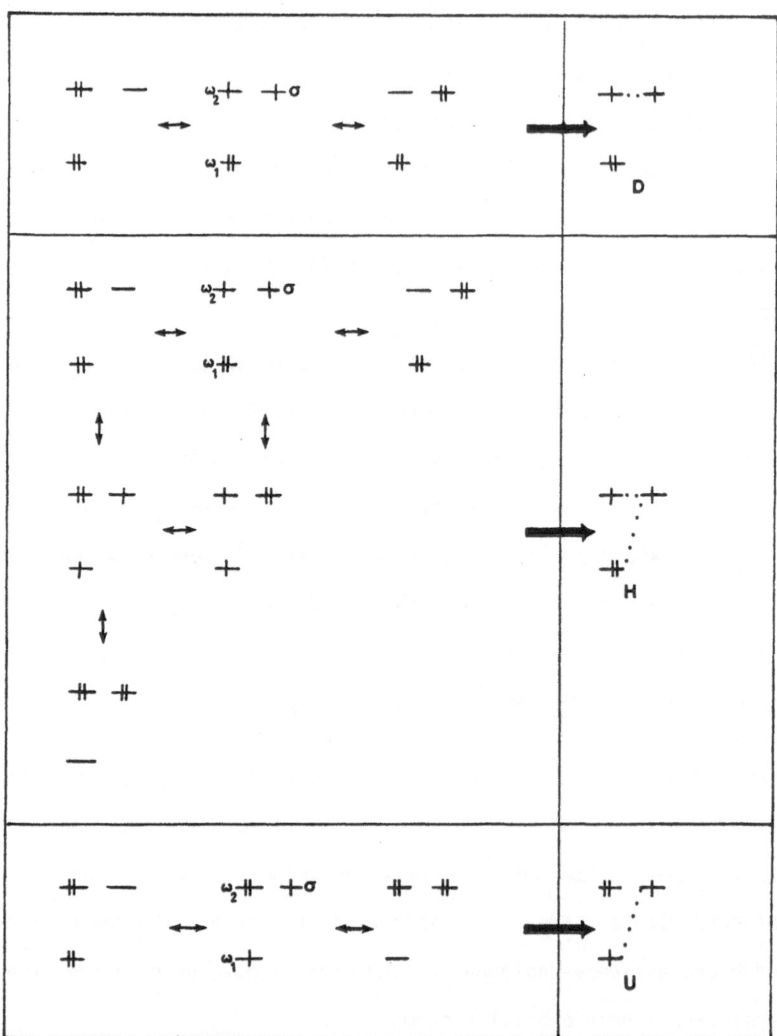

FIGURE 4: D. H, and U bond diagrams for three possible geometries of $\ddot{C}$-L imposing three different types of orbital symmetry constraints. Each bond diagram on the right is the representative of the manifold of interacting CW's on the left. Note that, in D, the CW interaction is due to the interaction (overlap) of ω_2 and σ, in U, it is due to the interaction (overlap) of ω_1 and σ, and, in H, it is due to the interactions (overlap) of ω_2 as well as ω_1 and σ. All CW's are assumed to interact exclusively by the CT mechanism.

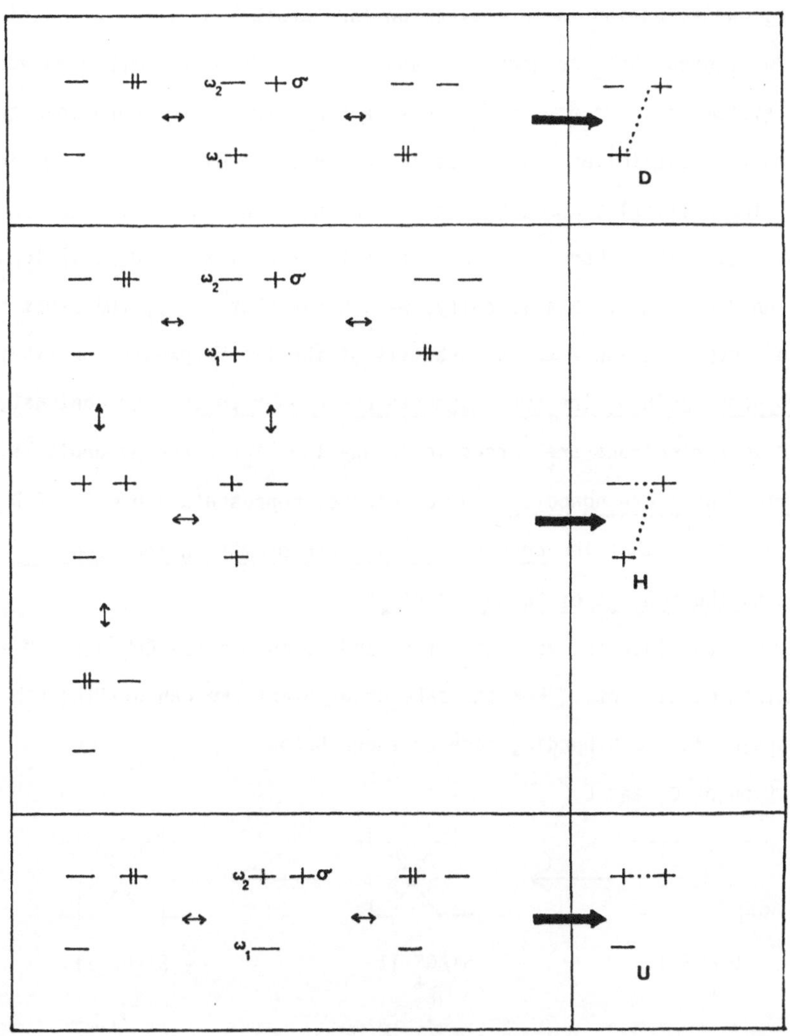

FIGURE 5: D, H, and U diagrams for three possible geometries of $\mathring{C}$-L imposing three different types of orbital symmetry constraints.

Turning our attention to four electron-four orbital systems, we consider the union of two closed shell fragments C: and L: to produce the doubly bonded composite system C=L. We now need twenty CW's and these are shown in Figure 6. Again, symmetry restrictions give rise to the three bonding modes shown in Figure 7. The critical V and I CW's are also shown in Figure 7. The reader should now notice that there are two different yet equivalent descriptions of the three bonding types. Specifically, we can say that, in D, the electrons are allowed to occupy the lower energy orbitals of the two fragments, or, that, in D, the two bonds joining the two fragments are highly polar. By contrast, in U, two of the four electrons are forced to occupy the higher energy orbitals, or, in U, the two bonds are nonpolar. H, of course, represents a hybrid of D and U. Accordingly, we see that the concept of "electron promotion and demotion" is equivalent to the concept of "bond ionicity".

Let us now consider the reaction of C: and L: to produce C=L under D and U stereochemical constraints. For the sake of argument, we can distinguish three reaction stages for each bonding mode as shown below.

I. D Reaction of C: and L:

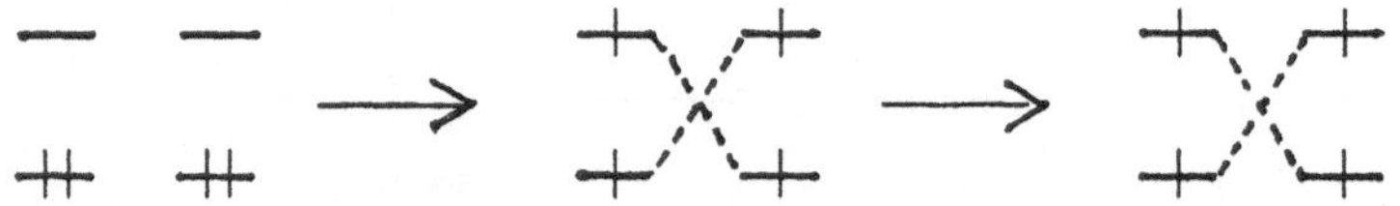

STAGE I D_R → STAGE II $D^{\ddagger}$ → STAGE III D_P

II. U Reaction of C: and L:

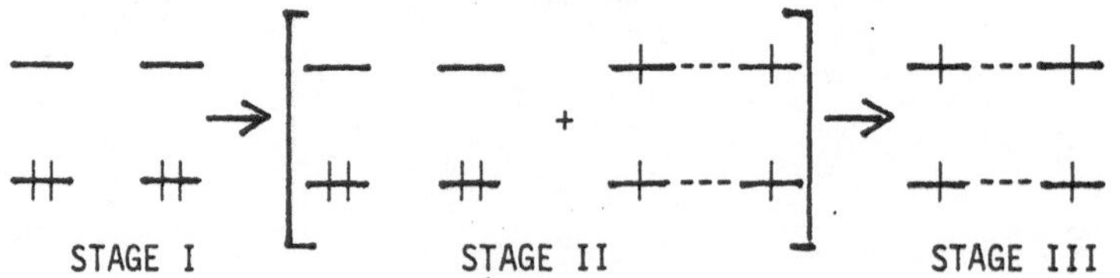

STAGE I U_R → STAGE II $U^{\ddagger} \alpha U_R - U_P$ → STAGE III U_P

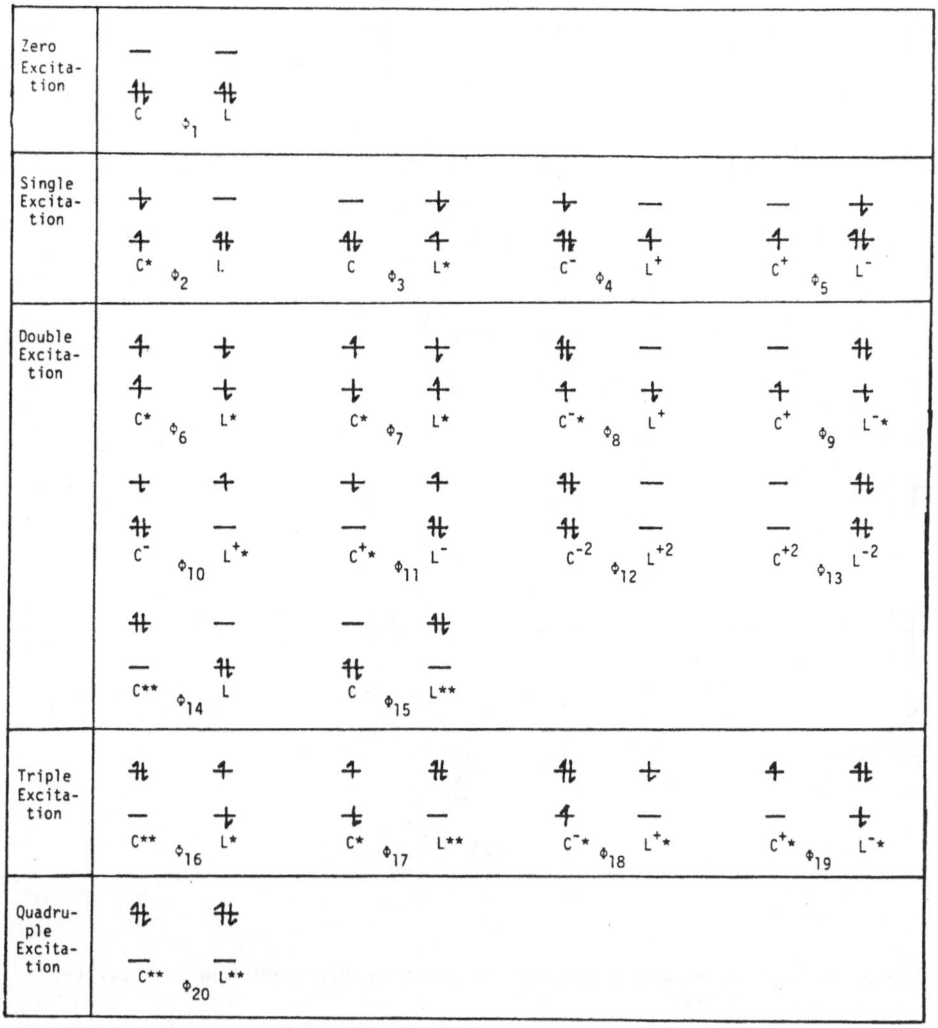

Zero Excitation	C Φ_1 L			
Single Excitation	C^* Φ_2 L	C Φ_3 L^*	C^- Φ_4 L^+	C^+ Φ_5 L^-
Double Excitation	C^* Φ_6 L^*	C^* Φ_7 L^*	C^{-*} Φ_8 L^+	C^+ Φ_9 L^{-*}
	C^- Φ_{10} L^{+*}	C^{+*} Φ_{11} L^-	C^{-2} Φ_{12} L^{+2}	C^{+2} Φ_{13} L^{-2}
	C^{**} Φ_{14} L	C Φ_{15} L^{**}		
Triple Excitation	C^{**} Φ_{16} L^*	C^* Φ_{17} L^{**}	C^{-*} Φ_{18} L^{+*}	C^{+*} Φ_{19} L^{-*}
Quadruple Excitation	C^{**} Φ_{20} L^{**}			

FIGURE 6: The twenty MOVB CW's required for the description of a four electron-four orbital system.

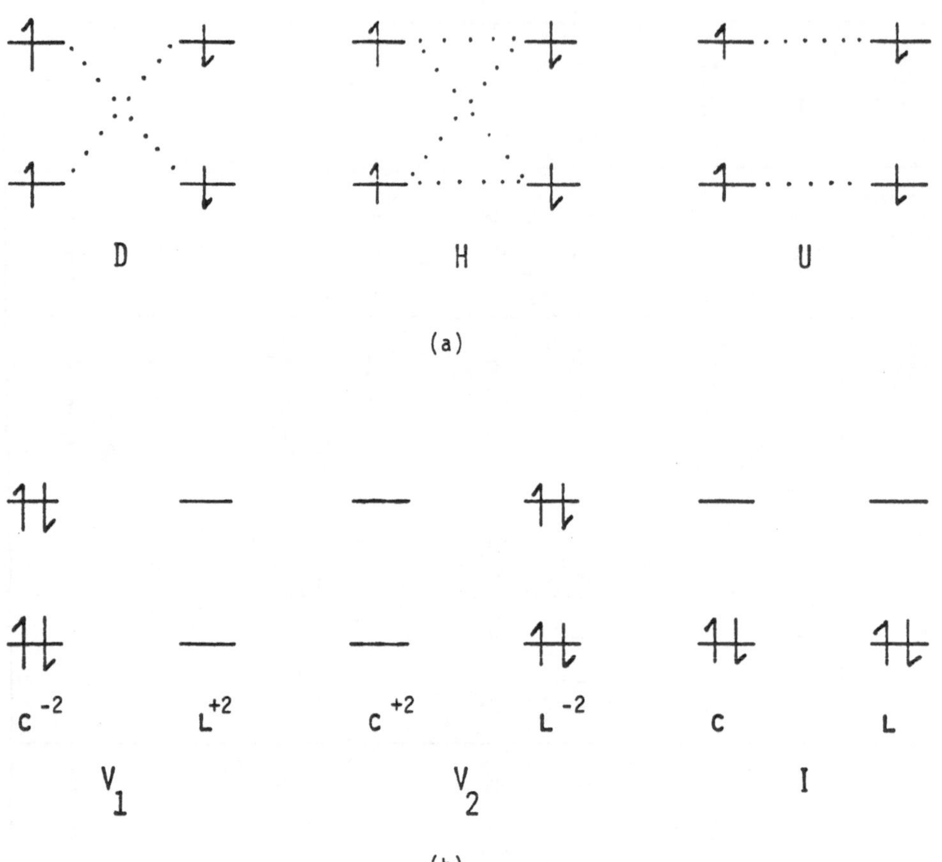

FIGURE 7: a) D, H, and U bonding in the four electron-four orbital C=L system.

b) V and I type CW's. As the energy of V_1 or V_2 decreases H bonding becomes increasingly favorable relative to D bonding. Conversely, as the energy of I decreases, H bonding becomes increasingly unfavorable relative to D bonding.

Stage I represents the reactant stage, stage II the reaction midpoint or transition state, and stage III the product stage. $U^{\ddagger}$ is the lowest singlet "diradical state",[27] and the bond diagrams depict how C and L become bound along the reaction coordinate. Because of the symmetry contraints, the D reaction is described by CW's Φ_1, Φ_4- Φ_7, Φ_{12}- Φ_{15} and Φ_{18}- Φ_{20} and the U reaction by CW's Φ_1, Φ_6- Φ_{15} and Φ_{20}. It is immediately obvious that the difference between the U and D reactions insofar as the transition state is concerned lies in the fact that D allows occupation of the low lying orbitals while U does so only in part being composed of a CW (Φ_1) which permits occupancy of the two lower energy orbitals and a set of CW's (implied by the bond diagram) which forces two electrons to occupy the two higher energy orbitals.

At the Extended Hückel theory level, U_R and U_P are degenerate at the reaction midpoint and they do not interact since bielectronic terms are neglected. Accordingly, either U_R or U_P is a good representation. This means that the difference between "allowed" and "forbidden" reactions lies in the nature of the bonds made at the transition states, with polar bonds made in the former and nonpolar ones in the latter. Now, many years ago Pauling suggested that the reaction shown below should be exothermic because combining four atoms in a manner which yields two nonpolar bonds is inferior to combining them in a way which leads to formation of two polar bonds because of the advantage of "ionic resonance" which is only significant in the latter case.[28]

$$A_2 + B_2 \rightarrow 2AB \qquad (45)$$

The conceptual identity of the two seemingly different problems discussed above is made clear by the bond diagrams shown below:

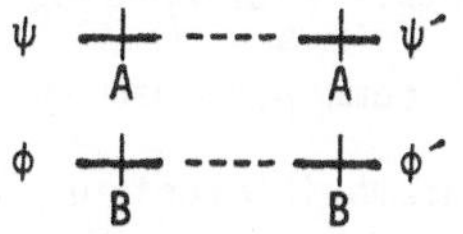

ψ A ---- A ψ´   ψ A ╲╱ A ψ´
φ B ---- B φ´   φ B ╱╲ B φ´

U 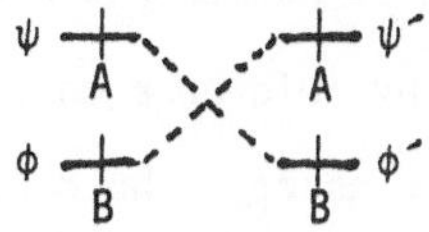   D

Clearly, $A_2 + B_2$ is entirely analogous to U_P, which is related to $U^{\ddagger}$, and 2AB entirely analogous to D_P, which is related to $D^{\ddagger}$. One could very well claim that the fundamental concept of "relative aromaticity", to which most chemists have been exposed since scientific infancy, is a special conceptual application of the above equation. Thus, MOVB theory has already allowed us to connect the MO theoretical idea of "relative aromaticity" with the well tested VB theoretical idea of "bond ionicity".[29]

The above discussion makes one thing clear: at the level of MOVB theory, chemical stereoselection is reflected in the relative energy of D- and U-bound systems and the characteristic difference between D and U bonding is that, while interfragmental overlap is kept nearly constant, the principal CW's describing the former have lower excitation energy, P_i, than the principal CW's describing the latter. Hence, we can say that the origin of chemical stereoselection in MOVB theory is the P_i terms of the energy matrix elements.

Two comments on bond diagrams are now in order. First, bond diagrams represent multicenter bonds. Thus, two center bonding becomes a partial case of multi-center bonding. Second, a detailed analysis of the CW composition of an MOVB wavefunction can be made possible by establishing the R CW as the universal frame of reference. The notation employed in the construction of MOVB bond diagrams projects this unambiguous choice of reference frame. Note that the dominant CW of a given wavefunction is not necessarily the reference R CW. The identity of the former depends on the energy interrelationship of the orbitals of the two fragments.

At this point, we open a parenthesis in order to comment on the differing languages of VB theory and MOVB theory based on the Core-Ligand dissection. As we have discussed before, VB theory of any type has a conceptual advantage over MO theory mostly because of the construct of the chemical "bond" or "antibond". Furthermore, it makes possible a physical interpretation of bonding and antibonding by reference to only three fundamental concepts, namely, the concept of excitation energy, the concept of classical coulomb interaction, and the concept of non- and semi-classical overlap interaction. These three concepts are embodied in the terms, P, G, and X, respectively, in terms of which the matrix elements of VB theory are expressed. Now, G plays a comparable role in both VB and MOVB theory. On the other hand, the roles of P and X are substantially different in the two formulations. Specifically, P controls only the magnitude of stereoselection in VB theory but determines stereoselection itself in MOVB theory. By contrast, X determines stereoselection itself in VB theory and controls only the magnitude of stereoselection in MOVB theory. Accordingly, while bielectronic overlap interaction can be neglected at both levels of theory as being merely the attenuator of monoelectronic overlap interaction, different approximations of the monoelectronic part of X are called for by the two theoretical approaches. In VB theory, stereoselection can enter via high order AO overlap terms. Thus, X is approximated by X' in VB theory but by X^{0} in MOVB theory, with the latter now being an approximate form of the former. The key point here is that the simplified expression of the overlap interaction which makes possible the qualitative application of MOVB theory to real systems is, in turn, made possible by the fact that stereoselection enters via the P rather than the X' term.

Finally, we wish to emphasize that the conceptual dichotomy identified above is not between VB and MOVB theory but rather between a VB theory defined over many fragments, i.e., atoms, and an MOVB theory defined over <u>two</u> fragments, i.e., the Core and Ligand fragments. In other words, it is the Core-Ligand dissection mode which is responsible for the considerable simplification of the VB energy expressions. These simplifications would not be feasible if we were to implement MOVB theory on <u>three or more</u> defined fragments. That is, if we were to dissect an ABCD system into three component fragments, e.g., AB+C+D, or four component fragments, A+B+C+D, we would have to implement MOVB theory in the former case and VB theory in the latter, using the equations of VB theory presented in the previous paper, if the fragment AO's overlap in a cyclic manner. It is the two-fragment dissection, e.g., AB+CD, which allows one to develop simple equations in the manner described in this paper.[30]

With a clear understanding of the subtleties of VB theory, in general, we can now proceed forward towards the realization of our goal: A "back of the envelope" MOVB qualitative theory of chemical bonding based on the Core-Ligand dissection.

D. The Representation of Elementary Systems

The systems with which we have dealt up to now can be labeled elementary systems as they contain few electrons in few orbitals. A more specific definition of the term "elementary system" is not useful because any large system can be viewed as a composite of elementary systems of arbitrary complexity. As we have seen, a ground elementary system can be represented by a bond diagram which constitutes the pictorial depiction of the corresponding optimal ground state wavefunction, Ψ. We now seek an approximate representation which makes evident how bonds are made in an elementary system which is more complex than those we have already discussed.

We illustrate our approach by reference to the specific case of a four electron-three orbital H-bound system such as the one shown below.

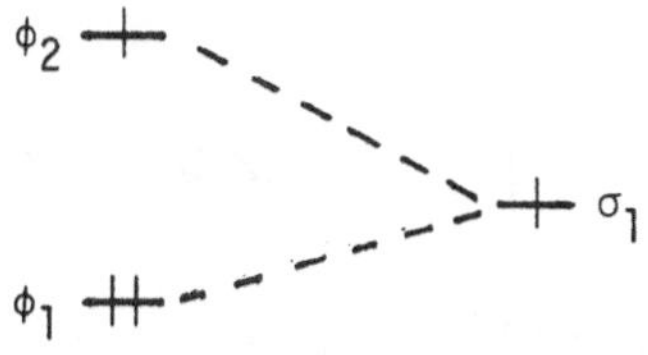

ϕ_2

σ_1

ϕ_1

Ω_H

We imagine the following situations:

a) The $\langle\phi_1|\sigma_1\rangle$ overlap integral is zero by symmetry while the $\langle\phi_2|\sigma_1\rangle$ one is not. In this case, we have D bonding as represented in Figure 8.

b) The $\langle\phi_1|\sigma_1\rangle$ overlap integral begins to increase as the $\langle\phi_2|\sigma_1\rangle$ one begins to decrease due to relaxation of symmetry constraints. In such a case we have H bonding. The wavefunction of this H-bound system can be described in the way illustrated in Figure 8, i.e., the H-bound system can be represented by a resonance hybrid of a "U" and a "D" bound system with the wiggly arrow informing us that the "U" and "D" bond diagrams contain a common CW and that an additional CW must be added to the set for completeness. These CW's are the ones shown below.

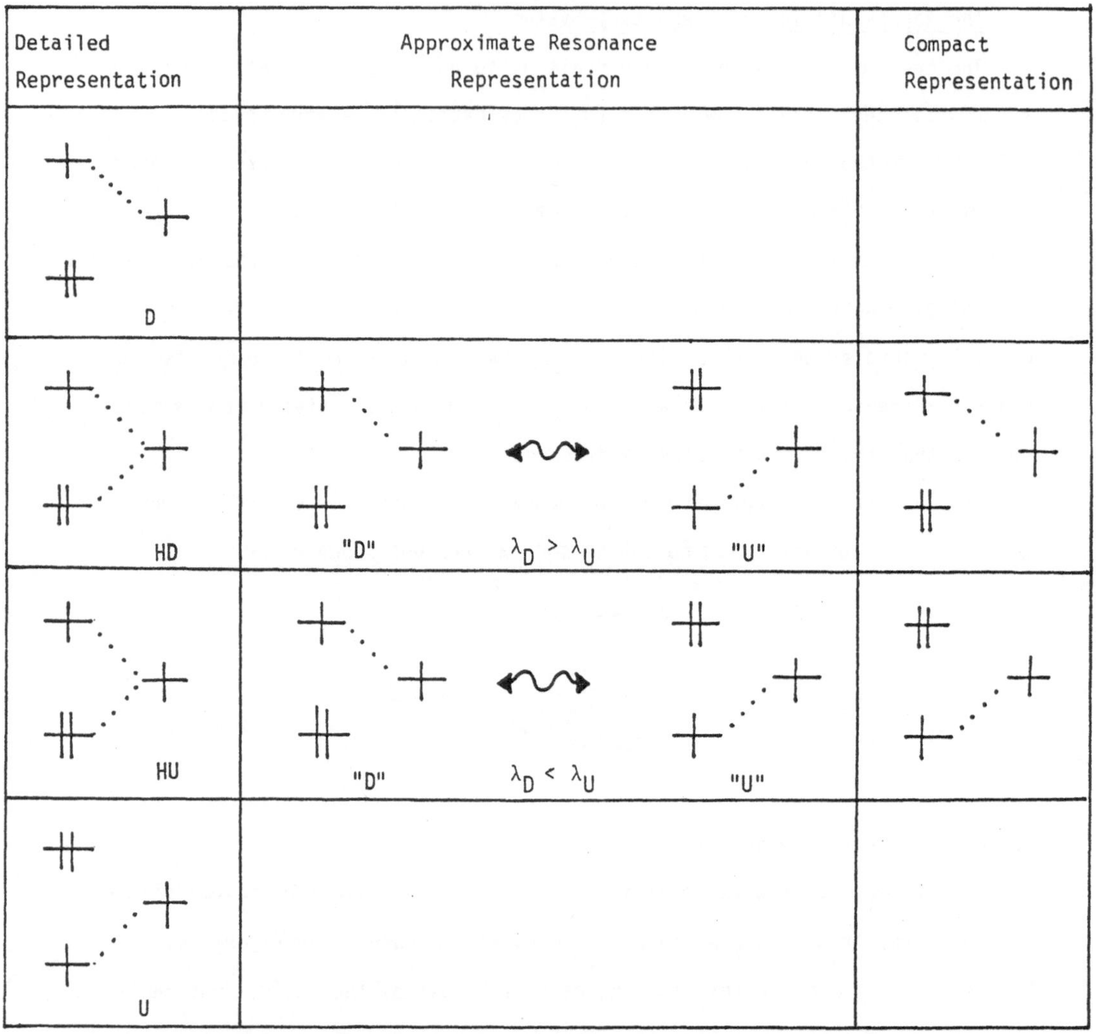

Detailed Representation	Approximate Resonance Representation			Compact Representation
D				
HD	"D"	$\lambda_D > \lambda_U$	"U"	
HU	"D"	$\lambda_D < \lambda_U$	"U"	
U				

FIGURE 8: Diagrammatic representation of U-, HU-, HD-, and D-bound 3 orbital - 4 electron systems. The compact representation of HU and HD bound systems reflects the resonance contributor ("U" or "D") of major importance.

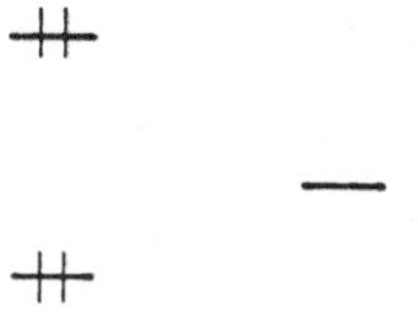

Common CW 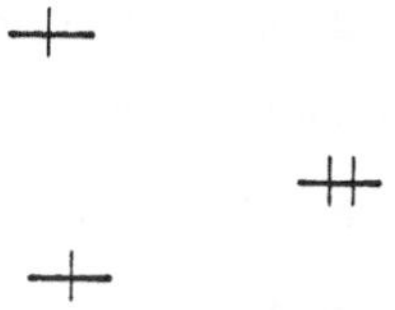   Additional CW

In describing the H-bound system, coefficients or weight factors, λ_U and λ_D, must be included as multipliers of the "U" and "D" bond diagrams, respectively. Accordingly, we may write

$$\Omega_H' = \lambda_D \text{"D"} + \lambda_U \text{"U"} \tag{46}$$

This is an approximate wavefunction and it is instructive to examine in detail how it differs from the true wavefunction of the H-bound system.

The variationally determined wavefunction describing the ground state of the four electron-three orbital system, Ω_H, is pictorially represented by the bond diagram shown above. Using the CW nomenclature introduced before, Ω_H can be written as follows:

$$\Omega_H = c_1 I + c_2 R + c_3 V + c_4 R^* + c_5 I^{**} + c_6 P \tag{47}$$

The variationally determined "U" and "D" wavefunctions are:

$$\text{"D"} = d_1 I + d_2 R + d_3' V \tag{43}$$

$$\text{"U"} = d_3'' V + d_4 R^* + d_5 I^{**} \tag{49}$$

The approximate wavefunction, Ω_H', is then:

$$\Omega_H' = \lambda_D(d_1 I + d_2 R + d_3' V) + \lambda_U(d_3'' V + d_4 R^* + d_5 I^{**}) \tag{50}$$

Now, Ω_H' differs from Ω_H insofar as it has been obtained from the latter by assigning $c_6=0$ and modifying each c_i by δc_i so that

$$c_1 = \lambda_D d_1 + \delta c_1 \tag{51a}$$

$$c_2 = \lambda_D d_2 + \delta c_2 \tag{51b}$$

etc.

The values of λ_D and λ_U can be determined from the condition that the sum of the absolute values of the δc_i's attains a minimum. A renormalization of the wavefunction, Ω', produced in the manner described above completes the procedure. It is evident that if the deleted CW's have high energy relative to some or most of the undeleted ones, i.e., if the coefficient, c_i, of a deleted CW, Φ_i, tends to zero in the true wavefunction, Ω, then the bonding of the system described by Ω can be conveniently analyzed in terms of the elements, i.e., "resonance bond diagrams", of the approximate wavefunction, Ω', which has been constructed by truncation of Ω. Note the key role of the truncation: To produce an approximate wavefunction which is totally expandable in terms of elements having clear chemical and physical significance. Henceforth, we shall denote the procedure outlined above as the Truncation procedure. It constitutes a device for representing approximately the correct, variationally determined MOVB wavefunction in terms of elements which have physical and chemical significance. We shall use this approach repeatedly in order to develop an understanding of the physical and chemical meaning of the MOVB wavefunctions of complex (as opposed to model) molecular systems.

Next, we distinguish the following possibilities and obvious relationships:

1. If $\langle\phi_2|\sigma_1\rangle > \langle\phi_1|\sigma_1\rangle$ then $\lambda_D > \lambda_U$.
2. If $\langle\phi_2|\sigma_1\rangle = \langle\phi_1|\sigma_1\rangle$ then again $\lambda_D > \lambda_U$ because of the smaller excitation energy involved in D bonding.
3. If $\langle\phi_2|\sigma_1\rangle < \langle\phi_1|\sigma_1\rangle$ then a point is reached where $\lambda_U > \lambda_D$.

When $\lambda_D > \lambda_U$, we say that the system is HD-bound, i.e., that it is H-bound in a way which makes it resemble a D-bound system. When $\lambda_D < \lambda_U$, we say that the system is HU-bound. Accordingly, we convene that the compact representation of an H-bound system is such that it reflects the contributor "U" or "D" which is of major importance. These conventions are illustrated in Figure 8. Henceforth, contributor bond diagrams will be referred to as resonance bond diagrams.

c) The $\langle\phi_2|\sigma_1\rangle$ overlap integral is zero by symmetry but the $\langle\phi_1|\sigma_1\rangle$ one is not. In this case, we have U bonding (Figure 8).

The above discussion makes clear that there is a bonding continuum:

$$D \rightarrow HD \rightarrow HU \rightarrow U$$

The application of MOVB theory to diverse stereochemical problems will be illustrated in the last part of this paper. In this initial phase of the presentation of the theory, we shall make use of the detailed representation of H-bound systems. This will be abandoned in subsequent applications described in following papers in favor of the compact representation. By then, it is hoped that the reader will be familiar with the "conceptual mechanics" of MOVB theory.

The methodology developed above can be used to describe any elementary system regardless of the degree of complexity. Thus, the wavefunction of a many electron - many orbital H-bound elementary system can be expressed as follows:

$$\Omega'_H = \lambda_1 \Xi_1 + \lambda_2 \Xi_2 + \ldots + \lambda_k \Xi_k \qquad (52)$$

The Ξ_k 's are resonance bond diagrams and Ω'_H is constructed in the way described above. The resonance bond diagram which places most electrons in the lower energy orbital, say Ξ_1, will be denoted by "D" and the one which places most electrons in the higher energy orbitals, say Ξ_2, will be denoted by "U". Thus,we can write

$$\Omega'_H = \lambda_D \text{"D"} + \lambda_U \text{"U"} + \lambda_3 \Xi_3 + \ldots + \lambda_k \Xi_k \qquad (53)$$

The principal resonance diagram may be "D", "U",or Ξ_k depending on the situation at hand.

Finally, according to the analysis presented above, D and U bonding can be thought of as limiting forms of H bonding. Thus, for example, Ω_D and Ω_U can be thought of as extreme cases of Ω_H. Hence, any system is adequately represented by Ω_H. On this basis, one may drop the subscript H with the understanding that Ω may represent any kind of bonding depending on the situation at hand. The same is true for the corresponding approximate forms.

The concepts which we have developed up to now pertain to elementary systems. Our next goal will be to find a way to represent complicated molecules in such a way that the corresponding total wavefunctions are expressible in terms of elementary system wavefunctions or related constructs, so that we can make use of the concepts we have already developed. This problem will be taken up after we take a closer look at the relative merits of U, H,and D bonding and their dependence on the electronic structure of the core and ligand fragments.

E. The Basic Stereochemical Problems

In qualitative theory of the FO-PMO type, simplifying approximations have led to a rather simplistic view of highly complex problems. At the level of qualitative MOVB theory, complexity represents a welcome challenge. In preparing to meet this challenge, we must now recognize the different types of bonding situations one is apt to encounter in a wide variety of chemical problems and investigate them in detail. In order to do so, we must first examine the consequences of geometrical and constitutional changes on the orbital interactions within the substrate. What orbital interactions are triggered or suppressed as a linear molecule bends, as rotation about a bond takes place, as one geometric isomer is converted to the other, or, as the orientation of approach of one system towards another is changed? These are the questions which must be answered before we can formulate a general theory of the chemical bond.

As we have stated before, a molecule can be viewed as a system composed of a core and a set of ligands. We will be interested in the effect of a stereochemical change on the interactions of the symmetry orbitals of the core with the symmetry orbitals of the ligands. The latter are modified in a characteristic manner depending upon the acting stereochemical perturbation in a way which allows the following classification of elementary stereochemical problems:

a) Type I rearrangement. In Figure 9, we show an example of a stereochemical change, i.e., bending, which has the following result: The ligand orbital maintains essentially undiminished overlap with one core orbital while in addition overlapping with a second core orbital upon bending. Now, the linear arrangement is compatible with either D or U bonding, depending on the number of electrons, but the bent form is compatible only with H bonding. In this case, we convene that H will be symbolized by H' in order to denote the

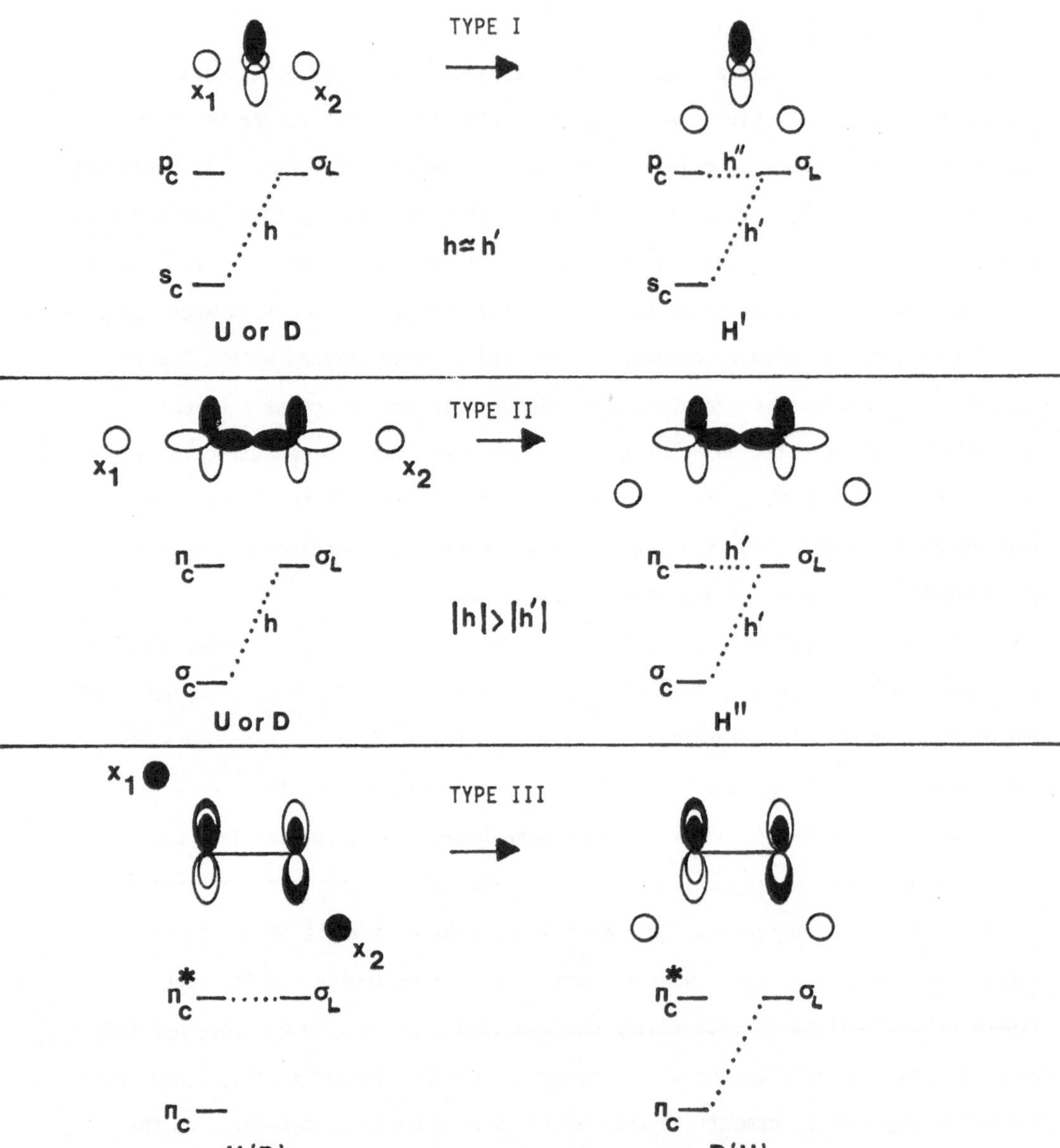

Figure 9: Three prototypical rebonding situations brought about by bending. x_1 and x_2 are AO's spanning the ligand fragment, L. The other AO's or MO's span the core, C. h symbolizes the resonance integral over MO's which is proportional to the corresponding overlap integral, s_{mn}. In Type I, hybridization occurs without impairing s_C-σ_L overlap. In Type II, hybridization occurs at the expense of σ_C-σ_L overlap. In Type III, a U(D) to D(U) transformation occurs by conservation of spatial overlap.

fact that spatial overlap is not reduced as a result of the stereochemical change which transformed a U or D system to a hybridized one. A U or D to H' conversion is expected in any stereochemical change which brings about the indirect mixing of two orbitals of one fragment, one of which has spherical symmetry and one which does not, via interaction with an orbital of the other fragment brought about by the stereochemical perturbation.

b) Type II rearrangement. In Figure 9, we show an example of a stereochemical change, i.e., bending, which has the following result: The ligand orbital σ_L loses overlap with the core orbital σ_C while gaining overlap with the π_C core orbital. Once again, the linear arrangement is compatible with D or U bonding and the bent form with H bonding. In this case, we convene that H will be symbolized by H'' in order to denote the fact that spatial overlap is reduced as a result of the stereochemical change which transforms a U or D system to a hybridized one. A U or D to H'' conversion is expected in any stereochemical change which results in the indirect mixing of two nonspherical orbitals of one fragment via the interaction with an orbital of the other fragment.

Interestingly, Type II rearrangement typifies the classical problem of molecular complex dissociation exemplified by the transformation of H_3^- to H_2 plus H^-. In this case, the "middle" hydrogen plays the role of the core and the outer two hydrogens play the role of the ligands. This is schematically illustrated in Figure 10.

The example of Figure 10 brings to mind an important warning which must be heeded by those who wish to use MOVB theory as a working theoretical tool. Specifically, the concepts of MOVB theory are developed over a basis of symmetry adapted fragment AO's and MO's. When two symmetry adapted fragment MO's are degenerate, one may arbitrarily replace them by their linear combinations. If, as a result of this replacement, the orbital basis is transformed into an AO

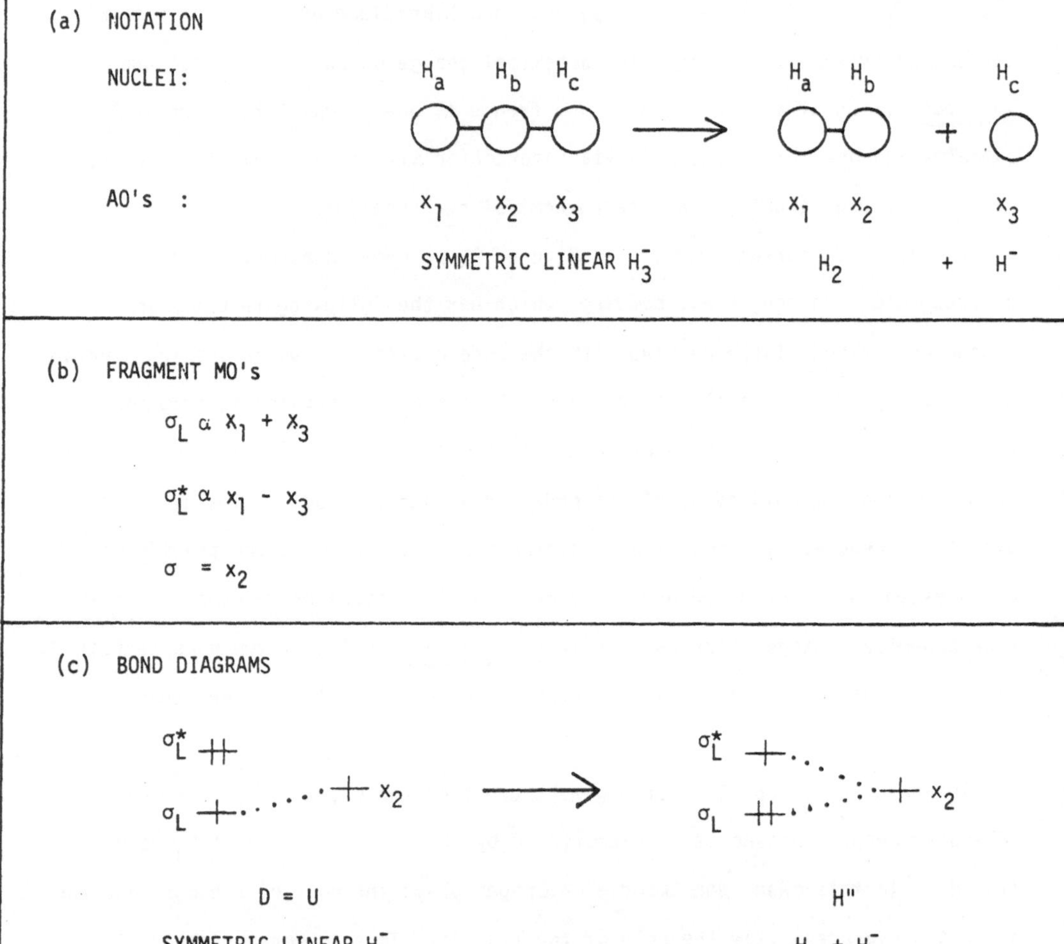

FIGURE 10: The construction of bond diagrams for the description of the $H_3^- \rightarrow H_2 + H^-$ dissociation process. H_a and H_c play the role of "ligands" and H_b the role of "core" throughout the reaction.

basis, then such an orbital basis becomes compatible with VB rather than MOVB concepts. In such a case, the labels U, D, H', and H'' lose their meaning.

c) Type III rearrangement. In Figure 9, we show an example of a stereochemical change in which the spatial overlap relationships of the two fragments are kept constant. This transformation is of the U to D type, or, vice versa.

The three categories defined above encompass the vast majority of the possible stereochemical problems which one is likely to encounter.

As we shall see, every molecule can be viewed as a sum of m electron-n orbital subsystems and each chemical reaction, in the broad sense of the word, as the process in which rearrangement occurs in every subsystem. For example, if we know that a reactant is made up of one U, one H', and one D subsystem which are transformed to D, D, and H'' subsystems, respectively, as a result of "reaction", we could predict the feasibility of the "reaction" itself if we had some way of evaluating the relative energetics of D, H (a collective term for H' and H''), and U bonding. In other words, we need a set of rules for predicting the energetic consequences of rearrangement as well as their variation as a function of the electronic nature of the constituent fragments of the "reactive substrate".

How is the electronic nature of fragments C and L, which comprise a target substrate, expected to influence the relative energies of the U, H, and D bonding in the substrate itself? Clearly, as C and/or L are varied, the relative energies of the fragment orbitals change. This, in turn, changes the relative energies of the basis CW's. Since each bond type involves a different set of interacting CW's, a variation of the electronic nature of C and/or L will affect U, H, and D bonding differently. We can then begin to see a rather simple scenario for the solution of a complex problem: *The direction of CT will determine which CW's are to be assigned the key roles and these will, in turn, control the relative energetics of the U, H, and D bonding modes.*

We now spell out the procedure for the development of general rules for optimal chemical bonding, i.e., theoretical guidelines for the qualitative prediction of the dependence of the relative energies of the U, H, and D bonding modes on the electronic nature of the fragments.

a) We confine our attention to a four electron-three orbital system $\ddot{C}$-L and a two electron-three orbital system $\overset{o}{C}$-L, where the two dots indicate an electron pair and the open circle a vacant orbital on C. The former can be thought of as the product of union of $\ddot{C}$• and L•, and the latter as a composite system made up of $\overset{o}{C}$• and L•.

b) We investigate the variation of the relative energies of the U, H, and D modes across a series of two systems which enforce different *intrinsic* donor-acceptor interrelationships between the two fragments C and L (in the absence of orbital symmetry constraints). That is to say, within this series, the intrinsic direction of CT changes from one extreme to the other. The two chosen systems, each characterized by a different orbital energy pattern, are shown below and the associated CW manifolds are shown in Figure 11.

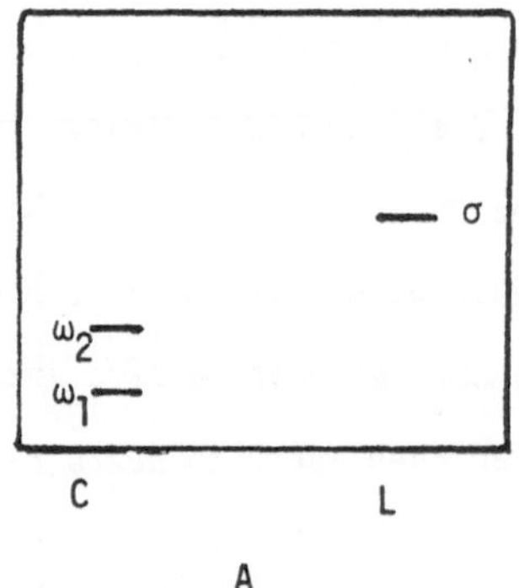

A

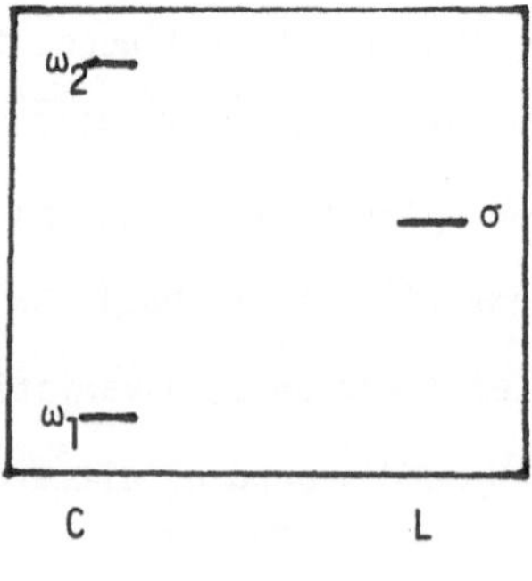

B

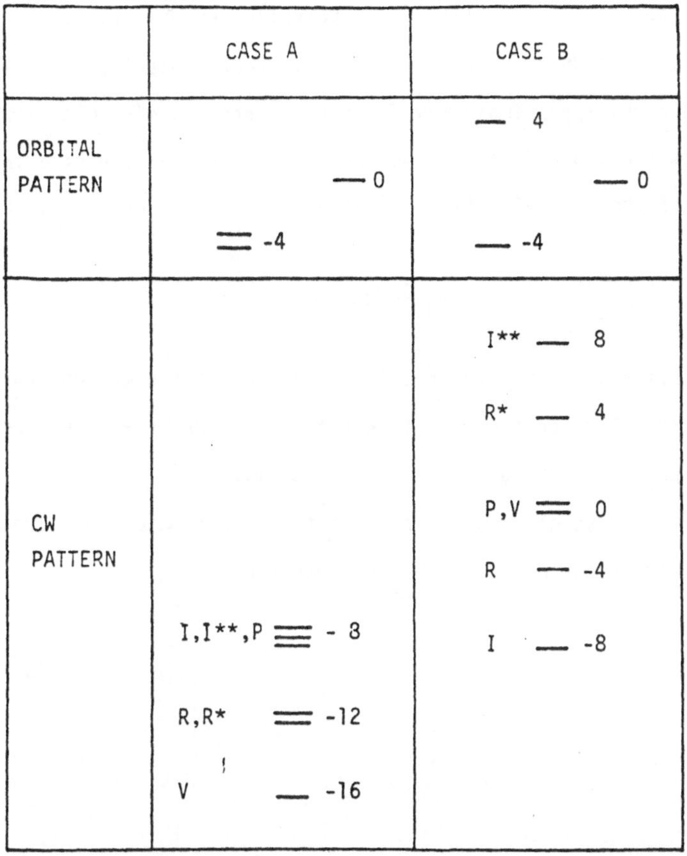

	CASE A	CASE B
ORBITAL PATTERN	— 0 ═ -4	— 4 — 0 — -4
CW PATTERN	I,I**,P ≡ - 8 R,R* ═ -12 V — -16	I** — 8 R* — 4 P,V ═ 0 R — -4 I — -8

FIGURE 11: The CW manifolds associated with two different orbital manifolds. In A, the CW's I** and R*, required for the description of U bonding, enter directly in the wavefunction of H through the first order mixing of R* with V. By contrast, this type of mixing can only be effected indirectly in B. The energy ranking of the CW's is according to their monoelectronic F_i's.

c) We compare bonding modes in a pairwise sense; U versus D, U versus H', U versus H'', D versus H', and D versus H''. Each comparison is illuminated by writing down the approximate wavefunction of each bonding mode for each of the two systems A and B and noting the trends. The convention used for the depiction of these wavefunctions is as shown below, where Φ_i is a CW.

$$\Phi_1 \leftrightarrow \Phi_2 \leftrightarrow \Phi_3 \leftrightarrow \text{etc.}$$

For qualitative purposes, the relative energies of the various CW's are evaluated by neglecting electron-electron interaction. Since we are interested in trends, this is a perfectly valid approach (Figure 11).

d) The results of computational tests are presented for comparison with the a priori predictions based on the MOVB wavefunctions.

F. Selection Rules for Chemical Bonding

1. Four Electron-Three Orbital Systems

The four possible modes of bonding in a four electron-three orbital system $\ddot{C}$-L are shown below.

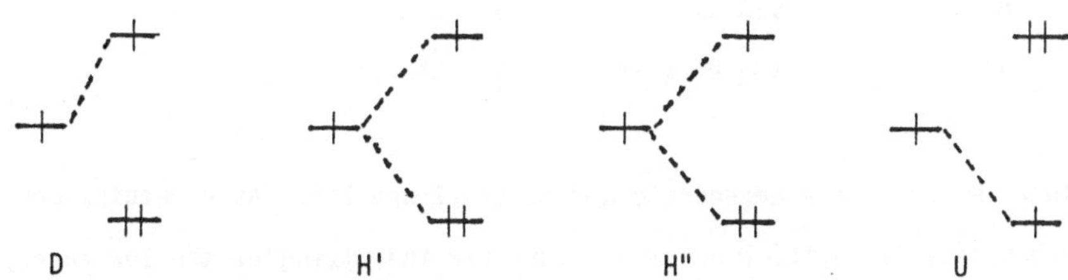

The make up of the wavefunctions of U, H', H'', and D is indicated below (see also Figure 1). For the purpose of discussion, it will be useful to differentiate the set of V , R, and I from the set of P, R*, and I** and refer to the former as the low energy, or, "low", set, and the latter as the high energy, or, "high", set.

U: V↔ R* ↔ I**

H': I↔ R↔ V↔ R*↔ I**↔ P

H'': I↔ R↔ V↔ R*↔ I**↔ P

D: I↔ R↔ V

1a. D versus U.

The wavefunctions of U and D for each of the two cases A and B are shown below with the principal contributor CW's underlined:

	CASE A	CASE B
D:	$\underline{V} \leftrightarrow R \leftrightarrow I$	$V \leftrightarrow R \leftrightarrow \underline{I}$
U:	$\underline{V} \leftrightarrow R^* \leftrightarrow I^{**}$	$\underline{V} \leftrightarrow R^* \leftrightarrow I^{**}$

In A, R and R* are degenerate and so are I and I**. As a result, D=U. As this degeneracy is split, D bonding occurs via the mixing of the low energy CW's V, R and I while U bonding is due to the mixing of V with the high energy CW's R* and I**. At the limit of B, the principal contributor of the wavefunction of D is I and this CW does not exist within the set of CW's responsible for U bonding. The principal contributor to the wavefunction of U is V. Since I has a much lower energy than V, D will attain energy significantly lower than U. We formulate the following rule: D is always more favorable energetically than U. As C becomes an increasingly better donor and L an increasingly better acceptor, the energetic preference for D over U will increase.

This rule can now be restated in a way which makes it independent of the choice of the fragments as follows: <u>The energetic preference for D over U increases as the I configuration attains the lowest energy relative to all other CW's.</u>

An important semantic distinction should now be made. Specifically, as we have already discussed, the concepts of "promotional energy control of optimal bonding" and "directional CT control of optimal bonding" are different formulations of one and the same principle. For example, the energy difference between U and D in the two systems C and C' shown below can be ascribed to either differential promotional energies or differential directional CT in the two systems.

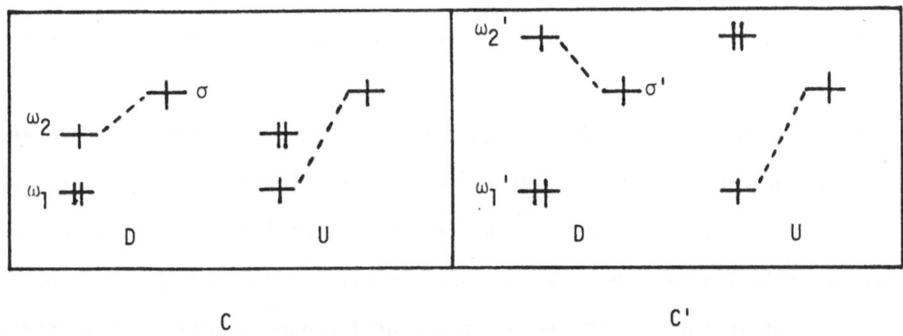

C C'

Specifically, we can say that the energetic preference for D over U is greater in C' relative to C because the $\omega_1' \rightarrow \omega_2'$ promotional energy is greater than the $\omega_1 \rightarrow \omega_2$ one. Alternatively, we can ascribe the same phenomenon to the fact that, in going from C to C' the I configuration is lowered in energy relative to the rest of the CW's. In short, either of the two physical descriptions of this one phenomenon is satisfactory. The important message here is that "excitation effects" and "electronegativity effects" are two sides of the same coin and not two different electronic mechanisms.

1b. D versus H'

The evolution of the wavefunctions of D and H' as one makes a transition from A to B is indicated below.

	Case A	Case B
D:	$\underline{V} \leftrightarrow R \leftrightarrow I$	$V \leftrightarrow R \leftrightarrow \underline{I}$
H':	$\underline{V} \leftrightarrow (R'+R'^{*}) \leftrightarrow (I'+I^{**}{}') \leftrightarrow P$	$V' \leftrightarrow R' \leftrightarrow \underline{I}' \leftrightarrow$ "High Set"

The following critical observations can be made with regards to the wavefunctions of H' and D:

i) The energy of the H' wavefunction depends on the energies of the "low" as well as "high" sets of CW's but the energy of the D wavefunction depends only on the energies of the "low" set.

ii) In terms of their principal contributors, the wavefunctions of H' and D <u>appear</u> similar along the entire range from A to B. However, there is one very important difference. Thus, in A, the principal contributor of H' and D is V, the energy of which is independent of bonding type. By contrast, in B the principal contributors of H' and D are I' and I, respectively, with I' having a much higher energy than I due to four-electron overlap repulsion absent in I because of symmetry constraints.

With the above considerations in mind, it is easy to understand exactly how the relative energies of H' and D will change as we make a transition from A to B. Specifically, in A the principal contributor V of H' and D cannot differentiate to first order between the two bonding modes. However, the effective mixing of the "high" with the "low" set of CW's, due to the fact that the six CW's are closely spaced in energy (Figure 11), renders H' more favorable than D on account of more extensive CW mixing. The situation changes dramatically in B where now the H' and D bonding modes are differentiated to first order by virtue of the fact that the leading contributor of the former is the higher energy I' and the leading contributor of the latter the lower energy I. In addition, the mixing of the "high" with the "low" set of CW's is no longer effective because the six CW's are no longer closely spaced in energy. Accordingly, D is now rendered more favorable than H' on account of overlap repulsion. This energy switchover between H' and D cannot be predicted at a level of theory which neglects overlap.

On the basis of this analysis, we can now formulate the following general rule: When V makes a dominant contribution to the D and H' wavefunctions, H' will lie below D energy. Conversely, if these conditions are not met the opposite will be true. These considerations are illustrated in Figure 12.

At this stage, it is important to develop some descriptive language which can be used constantly in future discussions. The proposed terminology is spelled out below and it is illustrated in Figure 13 by reference to the prototypical $\ddot{C}$-L system:

i) The bonding of every system is described by a bond diagram which shows the energy interrelationships of the orbitals of the fragments and defines the two-electron multicentric bonds joining the fragments.

ii) For every two-electron multicentric bond as defined by the bond diagram, the preferred direction of primary CT is denoted by a solid arrowhead as illustrated in Figures 13a and 13b. The result of primary CT is a CW which makes an important contribution to the total wavefunction.

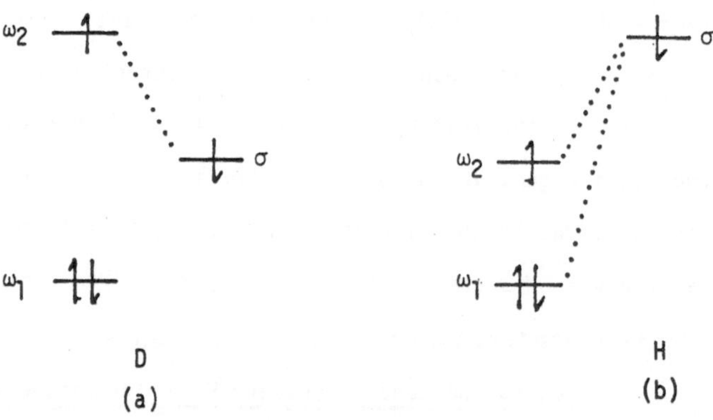

FIGURE 12: Selection rules for optimal bonding in a four electron-three orbital system. In (a), D bonding is favored because the I CW achieves low energy and the system adopts D bonding in order to avoid "first order" overlap repulsion. In (b), H bonding is favored because the V CW achieves low energy and, in the absence of "first order" overlap repulsion, the system adopts H bonding in order to benefit from extra delocalization.

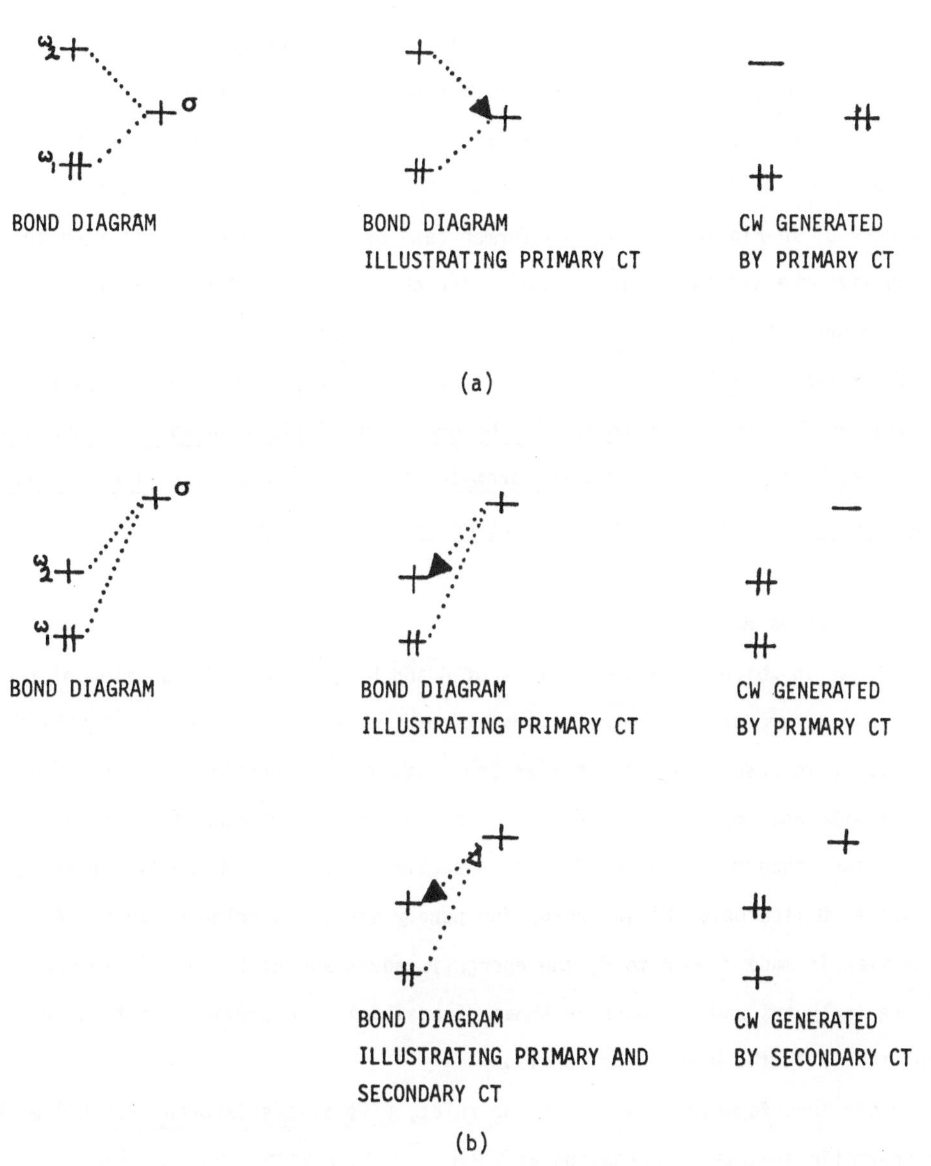

FIGURE 13: The pictorial definitions of primary and secondary CT in two different H systems, one in which the I CW makes a dominant contribution (a) and one in which the V CW is the lowest energy CW (b). Primary CT is indicated by black and secondary CT by white arrowhead.

iii) Depending upon the direction of primary CT, <u>secondary</u> CT may occur if there is a lone pair which can further delocalize in the hole created by primary CT. For example, primary CT renders secondary CT unfavorable and gives rise to overlap repulsion in Figure 13a. By contrast, primary CT sets the stage for secondary CT in Figure 13b. In the former case overlap repulsion will tend to destabilize the system, while in the latter case extensive delocalization will have the opposite effect.

The selection rule for H' over D preference can now be stated in a simple language as follows: <u>H' bonding will become increasingly favored over D bonding if primary CT occurs in a way which promotes further secondary CT, i.e., if it occurs in the direction of the fragment which carries a lone pair.</u>

1c. D versus H''

The way in which the wavefunctions of D and H'' change as a function of the electronic nature of the fragments and, in particular, as we make a transition from case A to case B, is entirely analogous to that encountered in the previous section with one important difference: The more strongly bound systems is now D rather than H'' because of loss of spatial overlap in H''. As a result, in case A, D lies below H'' in energy for purely spatial overlap reasons. As a transition is made from A to B, the energetic advantage of D over H'' increases for precisely the same reasons as those discussed in the previous section in connection with the D and H' comparison.

We can then formulate the following rule: <u>D is always favored over H'' with the energetic advantage decreasing as V makes an increasing contribution to the wavefunctions of D and H".</u> The same rule can be restated in more descriptive <u>language as follows: D bonding is favored over H" bonding with the energy</u>

difference decreasing as primary CT occurs in a way which promotes further secondary CT, i.e., as it occurs in the direction of the fragment which carries a lone pair. It is important to note that the energy differences E(D) - E(H') and E(D) - E(H'') have an identical dependence on the direction of primary CT. In other words, the dependence of E(D) - E(H) on the extent and direction of primary CT is spatial overlap invariant, i.e., it is the same regardless of whether H is H' or H''.

1d. U versus H'

This case need not be considered since the conclusions follow directly from the analyses of D versus U and D versus H'. Specifically, we predict that in case A, H' will lie below U and the energetic advantage of the former over the latter will increase progressively as the CW I' attains increasingly lower energy. Alternatively, we can say that the energetic advantage of H' over U will increase as primary CT occurs in a direction which prevents further secondary CT, i.e., it occurs in a direction away from the fragment which carries a lone pair.

1e. U versus H''

Again, this case need not be considered since the conclusions follow immediately from the analyses of D versus U and D versus H''. In case A, U will lie below H'' but as I'' becomes increasingly important or, equivalently, primary CT occurs in a direction which prevents further secondary CT, a point will be reached where there will be a switchover and U will end up higher than H''. Once again, it is important to note that the energy differences E(U) - E(H') and E(U) - E(H'') depend in the same way on the direction of primary CT, i.e., the dependence of E(U) - E(H) on the extent and direction of primary CT is spatial overlap invariant.

1f. Computational Tests

One of the contributions made in Part I was the delineation of equivalent theories. This, in turn, allows one to use MO computations in order to test VB theoretical constructs with knowledge of what brand of MO theory corresponds to which brand of VB or MOVB theory. In the case at hand, and because we are interested in ground state stereochemical problems, we can test the various predictions made above by Extended Hückel MO theoretical computations.[31] That is to say, we can approximate the MOVB theory as formalized in this paper by EHVB theory which is equivalent to EHMO theory and, thus, perform EHMO calculations to test our predictions. In a formal sense, this means that in performing the computational tests, we neglect the two-electron part of P and all of G in the equations of the diagonal and off diagonal matrix elements, the bielectronic matrix elements P_{ij} and W_{ij}, the polarization matrix element p_{ij}, and all two-electron parts of the overlap energy, X. In doing so, the test computations will reflect primarily CT interaction (i.e., d_{ij} and D_{ij} matrix elements) which, after all, forms the basis for the formulation of the concepts outlined before. When spatial overlap is large, as it is in ground state molecules in their equilibrium geometries, approximate EHMO computational tests are perfectly justified because CT interaction is dominant due to the large, in absolute magnitude, d_{ij} and D_{ij} matrix elements. Thus, although EHMO theory exaggerates electron delocalization because of the inherent approximations, the predicted trends are reliable. To put it crudely, relative overdelocalization at the level of EHMO theory parallels relative optimal delocalization at the level of MOVB theory.

The comments made above pertain to application of EHMO theory to ground state stereochemical problems. The luxury of simplification, i.e., implementation of EHMO, equivalent to EHVB and EHMOVB, rather than MOVB theory, is not permitted in the following cases:

i) In dealing with processes which involve significant variations of the "classical" interaction terms, G_i, e.g., chemical reactions. In such cases, even the trends predicted by EHMO theory may be wrong.

ii) In making absolute predictions regarding any process with the possible exception of transformations accompanied by significant deexcitation, i.e., U → D and U → H bonding changes.

It is then obvious that in applying MOVB theory, one must be cognizant of the types of problems which allow for the luxury of simplification, i.e., implementation of EHMO theory, and those which make application of MOVB theory mandatory. With this in mind, let us now see how the results of EHMO computations compare with the predictions of MOVB theory regarding the dependence of bond type preference on the electronic nature of fragments based on exclusive consideration of CT interaction.

The results of computational tests of the general rules outlined before are displayed in Figure 14. Several points deserve attention. First and foremost, the calculated trends are as predicted by the MOVB approach based on exclusive consideration of CT CW interaction. Secondly, the energy curve for H'' can be approximately produced by an upwards displacement of the energy curve for H'. This confirms our expectation that the E(H)-E(U) and E(H)-E(D) dependence on the direction of primary CT is independent of spatial overlap, i.e., it is the same whether H is H' or H''. Thus, in comparative studies of U, H, and D bonding, we need not specify the exact type of H bonding, i.e., H' or H'', if we are interested in the qualitative dependence of energy differences on the nature of the constituent fragments. Finally, it is interesting to note that Figure 14 projects a fundamental point which could have been fully anticipated in the absence of explicit test calculations. Specifically, the energy gap separating U and H' bonding modes is invariably much greater than that separating H' and D bonding modes when $\varepsilon_2-\varepsilon_1>2eV$, i.e., whenever the $\omega_2-\omega_1$ energy gap is either modest or large. This is understandable because a U to H' transformation represents deexcitation and it is fully expected to be accompanied by a significant energy gain when the orbitals defining the deexcitation are separated by either a modest or a large energy gap. By contrast, the H' to D transformation involves a trade-off between extra delocalization, present in the H' form and absent in the D form, and overlap repulsion, present in the H' but not in the D form. Furthermore, extra delocalization in the H' form can only be achieved by electron promotion to higher lying orbitals. As a result, the H' to D transformation can be accompanied by energy gain or loss depending upon the direction of primary CT,

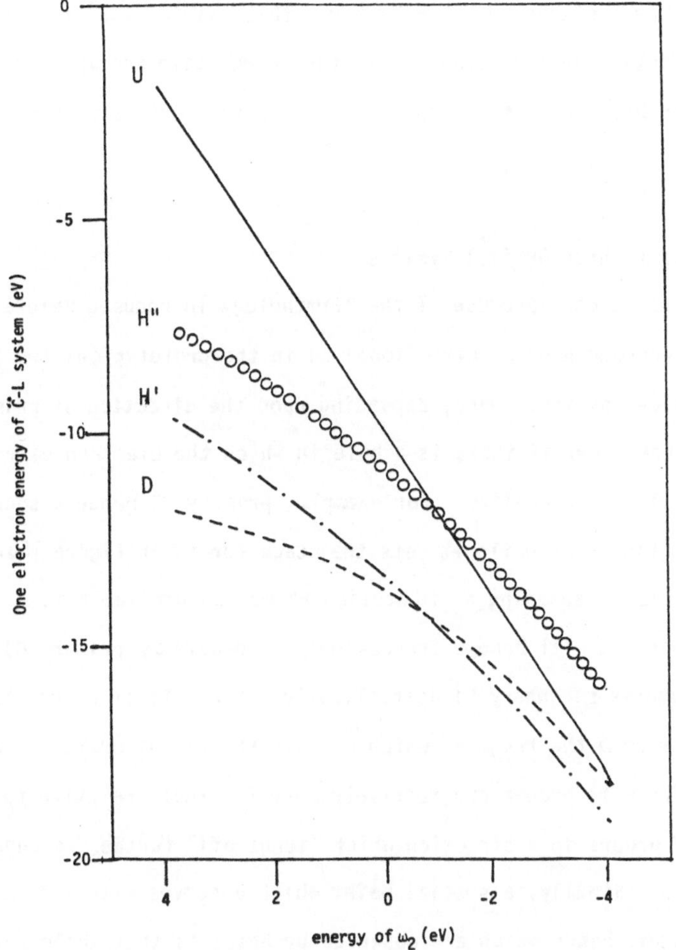

FIGURE 14: The one-electron energies of U, H", H', and D bonded $\ddot{C}$-L systems as a function of the energy of ω_2 when ω_1 and σ are fixed at -4 and 0 eV, respectively. As the value of ω_2 varies from -4 to +4 eV, the energy of the V CW increases while that of the I CW decreases, and vice versa. Note that the H" curve is obtainable by an upwards displacement of the H' curve. The values of the interaction matrix elements are as follows:

D: $\langle\omega_2|\hat{O}'|\sigma\rangle$ = 4eV

H': $\langle\omega_1|\hat{O}'|\sigma\rangle$ = 4eV and $\langle\omega_2|\hat{O}'|\sigma\rangle$ = 4eV

H": $\langle\omega_1|\hat{O}'|\sigma\rangle$ = 2eV and $\langle\omega_2|\hat{O}'|\sigma\rangle$ = 2eV

U: $\langle\omega_1|\hat{O}'|\sigma\rangle$ = 4eV

as we have already discussed. Furthermore, this gain or loss is relatively small in absolute magnitude compared to the energy gain accompanying a U to H' transformation because of the competition of "extra" delocalization and overlap repulsion.

2. Two Electron-Three Orbital Systems

Once again, we can make use of the terminology introduced before in order to describe the consequences of directional CT in the prototypical two electron-three orbital $\mathring{C}$-L system. Here, depending upon the direction of primary CT, secondary CT may occur if there is a hole in which the electron pair resulting from primary CT can delocalize. For example, primary CT renders secondary CT impossible in Figure 15a while it sets the stage for it in Figure 15b. By following the same reasoning as in Section F1 we can predict that H (H' or H'') bonding will become increasingly favorable as primary CT occurs in a way which promotes secondary CT delocalization, i.e., if it occurs in a direction away from the fragment which carries the vacant orbital. Also, H (H' or H'') bonding will become progressively more favorable relative to U bonding as primary CT occurs in a direction which "turns off" further secondary CT delocalization. Finally, a special point which deserves attention when dealing with multicentric bonds which are observed by holes is that while the direction and extent of primary CT may remain constant, the energy benefit incurred as a

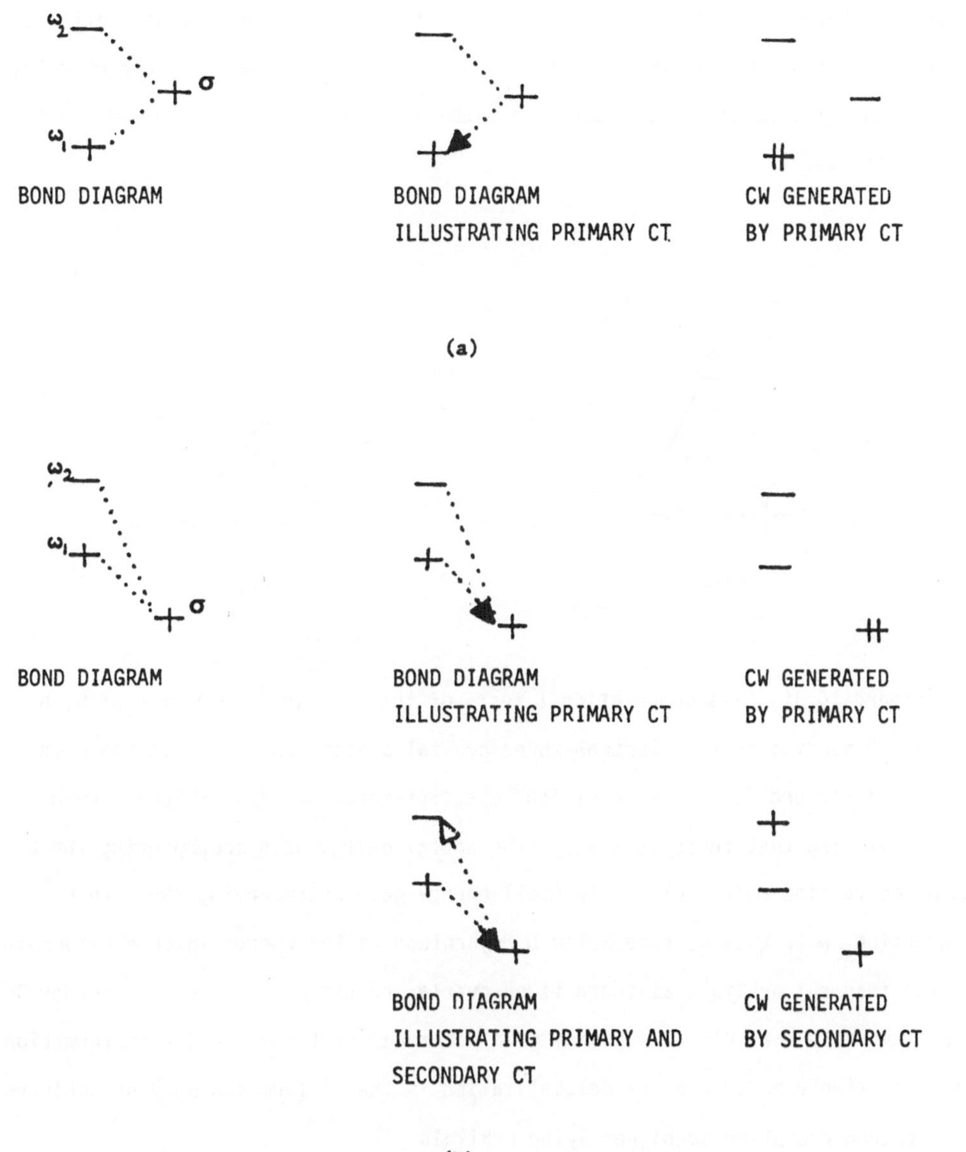

FIGURE 15: Primary and secondary CT is two different C̊-L systems. In (a) primary CT cannot set up secondary CT. In (b) primary CT sets the stage for secondary CT.

result of hybridization is now dependent on the energy of the vacant orbital in which the electron pair resulting from primary CT can delocalize. For example, the H mode is expected to be much more stabilized relative to the D mode in case F than in case E.

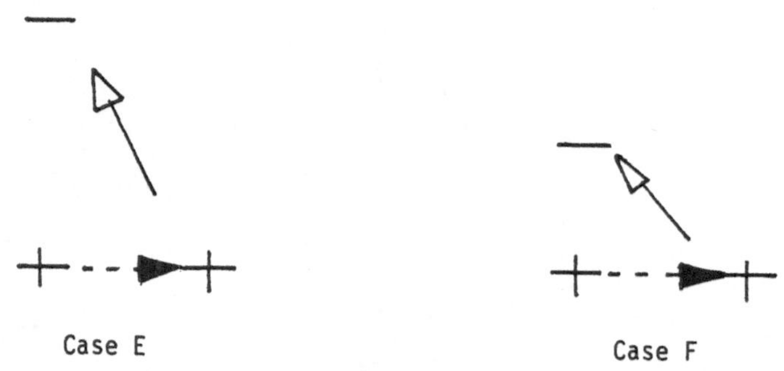

Figure 16 displays computational tests of the relative energetics of U, H', H'' and D bonding in two electron-three orbital systems analogous to the ones already described for the case of four electron-three orbital systems. Once again, we find that there is a very substantial energy gain accompanying the U to H' conversion and a relatively small energy gain accompanying the D to H' conversion. Note that H' lies below D regardless of the energy interrelationship of the fragment orbitals as there is no overlap repulsion component to render D more favorable than H'. The energy gain accompanying the D to H' transformation is small simply because extra delocalization in the H' form can only be achieved by electron promotion to higher lying orbitals.

In this and the previous sections, we have developed ideas, concepts, and predictions regarding the relative energies of the U-, H-, and D- bound systems which have general applicability. D is always better than U bonding, while H

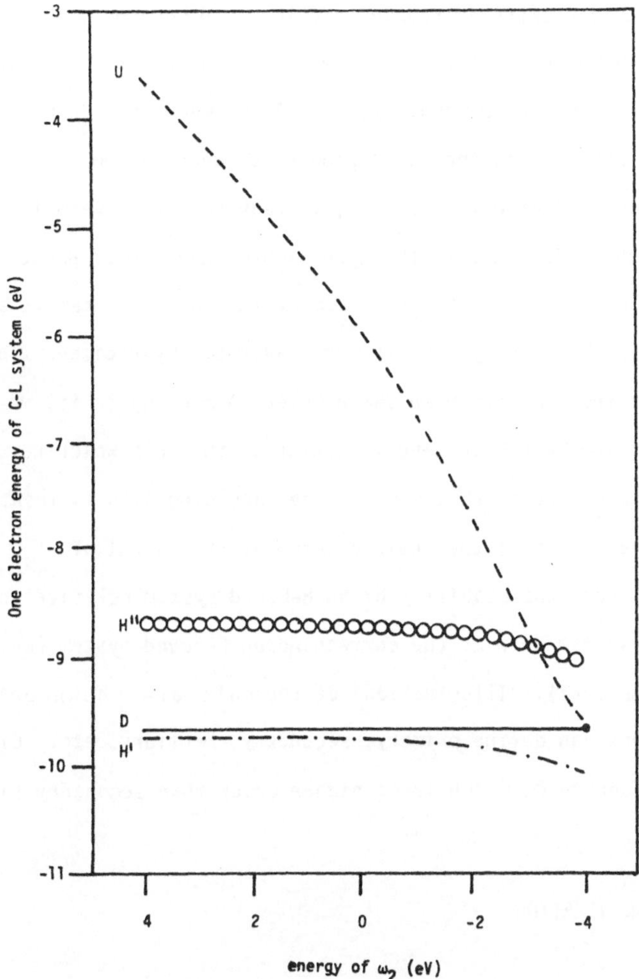

FIGURE 16: The one-electron energies of U, H", H', and D bonded C-L systems as a function of the energy of ω_2 when ω_1 and σ are fixed at -4 and 0 eV, respectively. The values of the interaction matrix elements are as follows:

D: $\langle\omega_1|\hat{O}'|\sigma\rangle$ = 4eV

H': $\langle\omega_1|\hat{O}'|\sigma\rangle$ = 4eV and $\langle\omega_2|\hat{O}'|\sigma\rangle$ = 4eV

H": $\langle\omega_1|\hat{O}'|\sigma\rangle$ = 2eV and $\langle\omega_2|\hat{O}'|\sigma\rangle$ = 2eV

U: $\langle\omega_2|\hat{O}'|\sigma\rangle$ = 4eV

bonding may achieve superiority or remain inferior to D bonding. The type of hybridization of an H bound system, i.e., good versus poor, can be defined by reference to the corresponding D-bound system. Thus, when an H-bound system tends to have lower energy than the corresponding D-bound system, we can speak of good hybridization, and vice versa. Now, in many problems, we will have to compare systems which differ only in the type of hybridization. Hence, we need some general conceptual device in order to ascertain, in a qualitative sense, which of two different H bound systems is more favorably hybridized. This conceptual device is the Hybridization Chain Rule. According to it, we depart from the perfect pairing CW (PP CW) and we construct the CW's which result from sequential one-electron hops so that none of the resulting CW's is identical to a preceding one. The length of the chain determines the extent of hybridization, and hence, the stability of an H-bound system relative to an arbitrary reference system such as the corresponding D-bound system (which may or may not be hypothetical). Illustrations of the rule are shown below. It is evident that one can define primary, secondary, tertiary, etc. CT. Henceforth, we will denote CT which is of higher order than secondary simply as higher order CT.

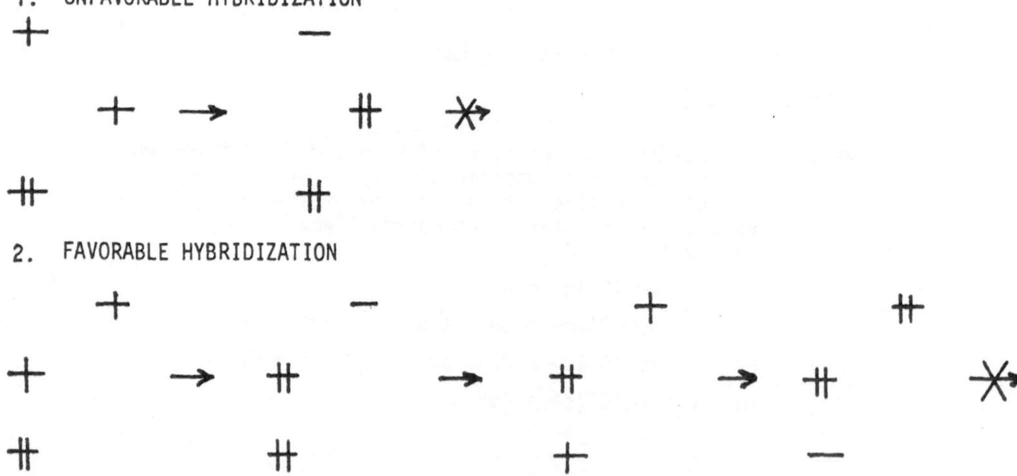

G. The Concept of the Effective Energy Gap

In our previous discussions of U, H, and D bonding modes, we have made the uniform assumption that the CT matrix elements are all equal. What happens when this approximation is not valid? For example, how does the picture we developed above change if the relative magnitudes of these matrix elements, h_1 and h_2, are as shown below?

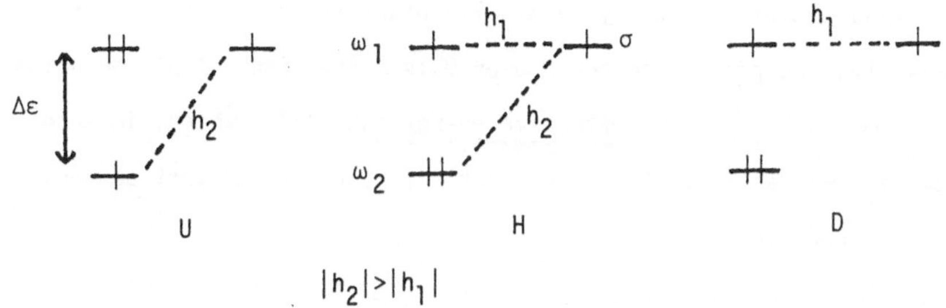

$|h_2|>|h_1|$

It will be sufficient to discuss only the U versus H case since the U versus D and D versus H comparisons can be made using similar reasoning.

As we have discussed before, the U→H conversion can be thought of as a deexcitation process which becomes increasingly favorable as I attains an increasingly lower energy relative to all other CW's. If the energy of the σ orbital is kept constant, this statement can be rephrased as follows: The energetic advantage of H over U increases as the energy gap, $\Delta\varepsilon$, separating the ω_1 and ω_2 orbitals increases, while h_1 and h_2 are kept constant. The question now becomes: How is the energetic preference for H over U affected if $\Delta\varepsilon$, h_1, and h_2 all change?

In order to answer the above question, we first define the quantities $\Delta\varepsilon$ and Δh as

$$\Delta\varepsilon = \varepsilon(\omega_2) - \varepsilon(\omega_1) \tag{54}$$

$$\Delta h = h_1 - h_2 \tag{55}$$

By reference to the bond diagrams shown above, we can determine that, if $\Delta\varepsilon$ is kept constant, bonding in U depends on h_2 while in H it depends on both h_1 and h_2. As h_2 increases in absolute magnitude relative to h_1 the bonding in H increasingly resembles that in U. That is to say, the H structure tends to keep the electron pair in the upper orbital ω_1 so as to make a strong bond as a result of the overlap of σ and ω_2. As a result, the energetic preference of H over U will diminish, and may eventually diappear, as Δh increases. It is then evident that the preference for H over U is a function not of the actual energy gap $\Delta\varepsilon$ but rather of some effective energy gap, $\Delta\varepsilon'$, which, in turn, depends on $\Delta\varepsilon$ as well as Δh. Now, the functional form of $\Delta\varepsilon'$ must be such that the following conditions are met:

a) When Δh equals zero, then $\Delta\varepsilon=\Delta\varepsilon'$

b) As Δh becomes increasingly negative the actual energy gap, $\Delta\varepsilon$, is magnified and $\Delta\varepsilon'$ increases.

c) As Δh becomes increasingly positive, the actual energy gap, $\Delta\varepsilon$, is reduced until it becomes equal to zero. Past this point, the effective energy gap becomes negative.

This functional form of $\Delta\varepsilon'$ is sketched below.

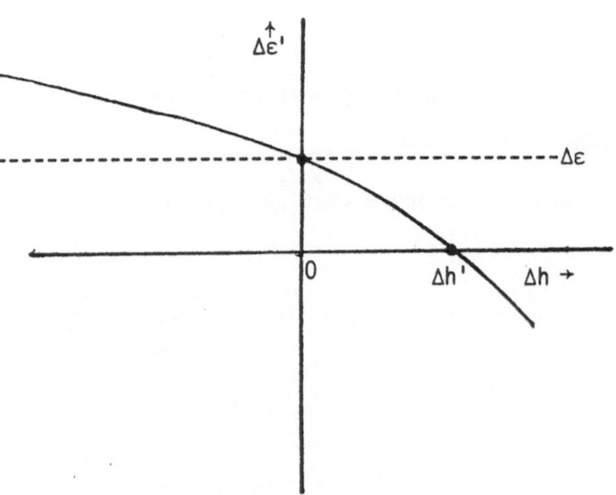

Recalling that H can be thought of as the resonance hybrid of "D" and "U", it is evident that, for values of Δh which are smaller than $\Delta h'$, "D" is the major contributor and the H system is of the HD variety. When $\Delta h = \Delta h'$, "D" and "U" make an equal contribution. Finally, when Δh is larger than $\Delta h'$, the major contributor is "U" and the H system is of the HU variety. This is exactly as demanded by the theory which envisions a continuum of bonding "flavors" ranging from D to H to U. We shall not attempt to derive the exact form of $\Delta\varepsilon'$ as this is unnecessary for qualitative treatments. Instead we shall simply define $\Delta\varepsilon'$ as follows:

$$\Delta\varepsilon' = F(\Delta\varepsilon, \Delta h)$$

For ages, chemists have wondered how atomic properties are effectively modified in molecules. A convincing solution of this problem can be offered on the basis of a "fragment in molecules" theoretical treatment. As we shall see, the properties of isolated atoms or molecules change dramatically when these atoms or molecules become a part of a molecule or supermolecule, respectively, and the failure to recognize this important fact is partly responsible for confusion and controversy in many areas of chemistry. In this paper, we have restricted our attention to effective energy gaps. In a following paper, we shall have the opportunity to see exactly how chemical phenomena can be misinterpreted if the analysis is based on consideration of atomic quantities (e.g., $\Delta\varepsilon$) rather than <u>effective</u> atomic quantities (e.g., $\Delta\varepsilon'$).

H. The Representation of Polyelectronic Systems

Since the original formulation of the theory of the chemical bond by Pauling and up until the present time, we have been conditioned to view ground states of molecules as a sum of two electron-two orbital bonds either implicitly or explicitly. As we have already seen, the abandonment of the concept of the two electron-two center bond in favor of the MOVB concept of the multicentric bond gives rise to a new formulation of the Woodward-Hoffmann rules according to which an "allowed" complex is characterized by D and a "forbidden" complex by U bonding. Now, up until this point, the illustrators of the principles of MOVB theory have been model systems, i.e., systems with few orbitals and few electrons. It is now time to ascend to a higher level of complexity for, after all, our goal is the development of a general, qualitative theory of chemical bonding of large systems. Specifically, we wish to answer the following key questions: Is it permissible to think of a complex molecular system as a composite of primitive "few electron-few orbital" systems in a way which renders the concepts developed before directly applicable? If this is not possible, what is the most suitable alternative option? The full dimensions of the problem and the solution which we propose can be illustrated by reference to the model four electron-four orbital system, A = A, depicted below. This is the simplest illustrator of a double bond comprised of one sigma and one pi bond, a well known feature of countless organic and inorganic molecules.

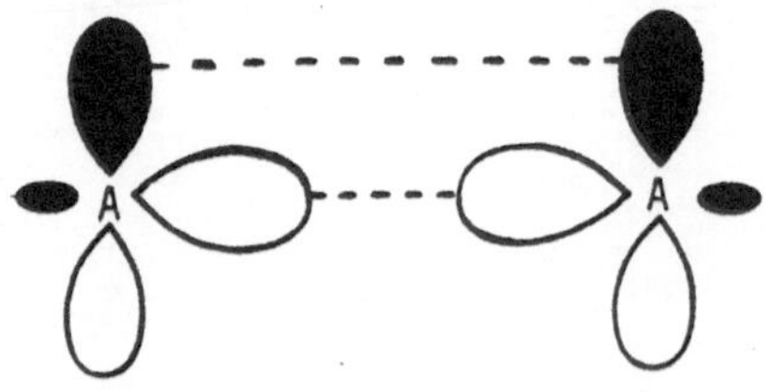

The electronic structure of A_2 can be described by the bond diagram shown below:

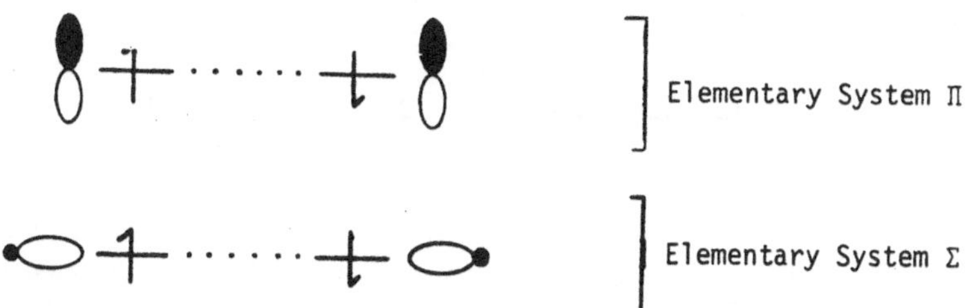

We now have two choices:

a) We can view A_2 as a single four electron-four orbital U-bound elementary system.

b) We can view A_2 as a composite of two elementary systems, Σ and Π, which cannot interact with each other in a one-electron sense.

Let us restrict our attention to the second alternative and ponder the question: Can we express the total energy of A_2 as a sum of the energies of Σ and Π?

Consider the Σ elementary system in the absence of the perturbing influence of the Π elementary system. The corresponding unperturbed sigma bond can be described as indicated in Figure 17a. Similarly, the unperturbed pi bond corresponding to the Π elementary system can be described as indicated in Figure 17b.

Finally, the composite system, A = A, in which there is a mutual perturbation of the sigma and pi bonds belonging to Σ and Π, respectively, can be described as indicated in Figure 18.

The "perfect" wavefunction of A = A is written as follows:

$$\Psi = \sum_i \xi_i Y_i \tag{57}$$

The energy of Ψ with respect to the complete Hamiltonian operator, H(σ,π), is E. Now, we can write an approximate product wavefunction as follows:

$$\Psi' = \Psi_\Sigma \cdot \Psi_\Pi \tag{58}$$

or,

$$\Psi' = \sum_a \sum_b \lambda_a \mu_b \Phi_a \cdot X_b \tag{59}$$

or,

ORBITALS: A — — A

$\Phi_1 \quad \leftrightarrow \quad \Phi_2 \quad \leftrightarrow \quad \Phi_3$

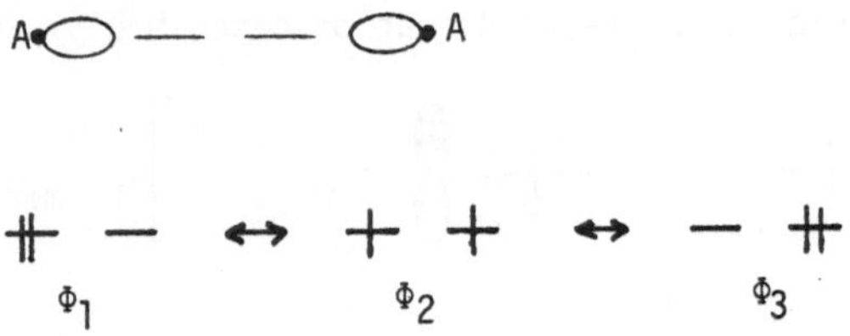

WAVEFUNCTION: $\Psi_\Sigma = \lambda_1\Phi_1 + \lambda_2\Phi_2 + \lambda_3\Phi_3$

(a)

ORBITALS: A — — A

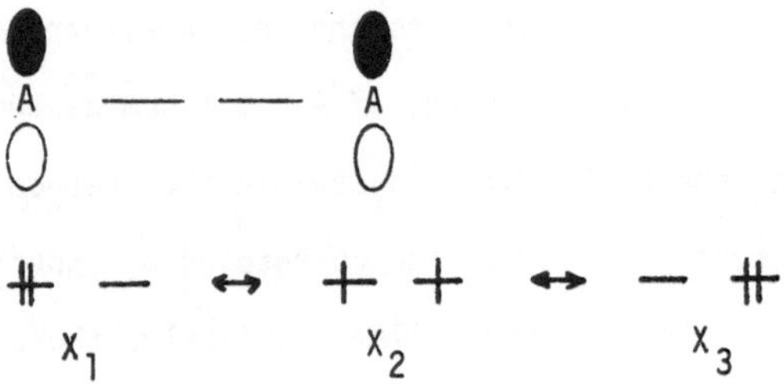

$X_1 \quad \leftrightarrow \quad X_2 \quad \leftrightarrow \quad X_3$

WAVEFUNCTION: $\Psi_\Pi = \mu_1X_1 + \mu_2X_2 + \mu_3X_3$

(b)

Figure 17. The independent Σ and Π wavefunctions.

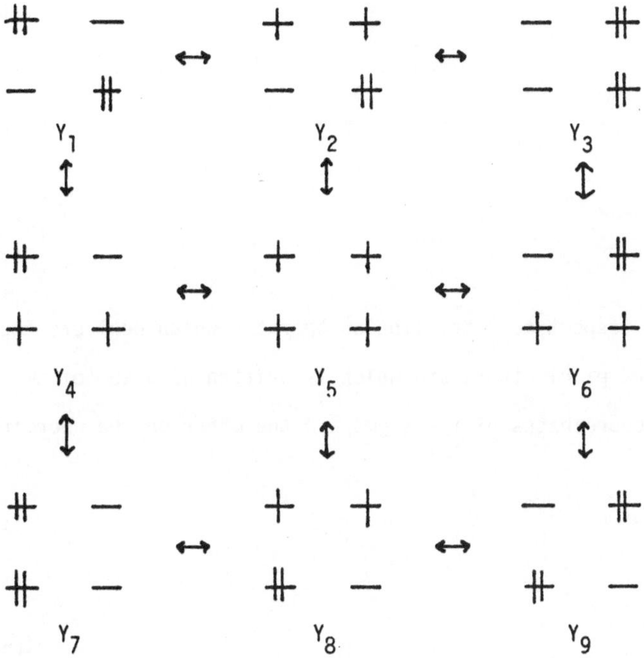

Y_1 Y_2 Y_3

Y_4 Y_5 Y_6

Y_7 Y_8 Y_9

$$\Psi = \sum_i \lambda_i Y_i$$

Figure 18. The total wavefunction, Ψ, of the composite A_2 system.

$$\Psi' = \sum_i \nu_i Y_i' \qquad (60)$$

where

$$\nu_i = \lambda_a \mu_b \qquad (61)$$

and

$$Y_i' = \Phi_a \cdot X_b \qquad (62)$$

The energy of Ψ' with respect to a Hamiltonian operator which neglects the interaction of the sigma and pi electrons and which is written as a sum of two parts, one depending on the coordinates of the sigma and the other on the coordinates of the pi electrons, is:

$$E' = E_{\Pi} + E_{\Sigma} \qquad (63)$$

if

$$\hat{H} = \hat{H}(\sigma) + \hat{H}(\pi) \qquad (64)$$

If we now replace ξ_i by ν_i in equation (57), we obtain:

$$\Psi'' = \sum_i \nu_i Y_i \qquad (65)$$

Recognizing that the energy difference between Y_i and Y_i' is equal to the interaction of the sigma and pi electrons, we obtain:

$$E'' = E_{\Sigma} + E_{\Pi} + \text{INT} \qquad (66)$$

where INT represents the overall sigma-pi interaction energy. Finally, the energy change of the A = A system as a result of a perturbation is:

$$\Delta E'' = \Delta E_{\Sigma} + \Delta E_{\Pi} + \Delta(\text{INT}) \qquad (67)$$

Clearly, if $\xi_i \simeq \nu_i$, then $\Delta E = \Delta E''$. If, in addition, $\Delta(\text{INT}) = 0$, we obtain:

$$\Delta E \simeq \Delta E_{\Sigma} + \Delta E_{\Pi} \qquad (68)$$

In most qualitative applications of MOVB theory, these approximations are justified. Special problems which can only be treated by non-approximate MOVB theory are discussed in a separate paper.

At this point, we digress briefly in order to point out that as long as the approximation $\Delta(INT) = 0$ is valid, qualitative MOVB theory is, in a sense, superior to monodeterminantal MO theory simply because at this level of theory the λ_a's and μ_b's are constrained. Hence, the ν_i's are constrained and Ψ can no longer describe properly the electronic reorganization attending the dissociation of the A = A double bond. As we shall see, this failure leads to incorrect conclusions in chemical problems in which a perturbation acts as to weaken the bonds connecting a core and a ligand fragment, in general.

A = A is a simple four electron-four orbital system. However, the concepts developed above are applicable to any composite system made up of elementary systems which cannot interact with each other in a one-electron sense, regardless of the complexity of these elementary systems. Thus, for example, a system such as the one represented by the bond diagram of Figure 19a can be treated exactly as the simpler A_2 system discussed above. It is then apparent that, in attempting to predict the energetic consequences of a geometrical change, we are entitled to view a composite system as a sum of elementary component systems, or, equivalently, as a sum of n electron-m orbital multicenter bonds, remembering the two approximations which paved the way to this vantage point.

On the basis of these considerations, one may construct an Independent Bond Model of electronic structure based on the following assertions:

a) Every complex system can be represented by a bond diagram. Alternatively, it can be represented by a linear combination of resonance bond diagrams as illustrated in Figure 19b.

b) Each bond diagram represents independent elementary subsystems symbolized as shown in Figure 19c.

c) The energy change of the total system due to a perturbation can be predicted by determining the effect of the perturbation on each elementary subsystem by

Subsystem 1 $\omega_3(\Gamma_b)$ .... $\sigma_3(\Gamma_b)$

$\omega_2(\Gamma_b)$ .... $\sigma_2(\Gamma_a)$

Subsystem 2

$\omega_1(\Gamma_a)$ .... $\sigma_1(\Gamma_a)$

(a)

Hybrid Representation

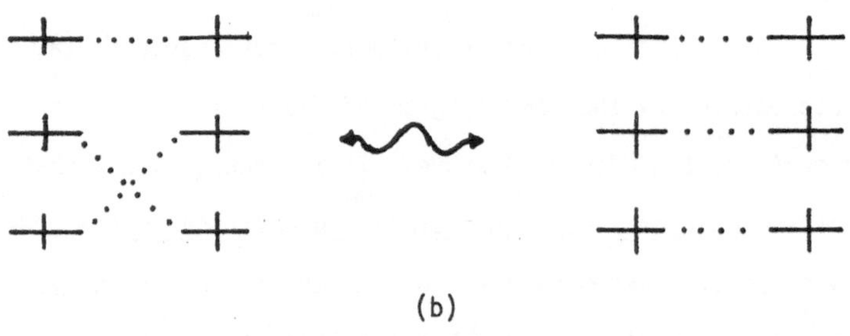

(b)

Subsystem Notation

$$\omega_1,\ \omega_2,\ \sigma_1,\ \sigma_2/4|\omega_3,\ \sigma_3/2$$

(c)

FIGURE 19 (a) Bond diagram for a six electron-six orbital A≡A system.
(b) Representation of A≡A as a resonance hybrid.
(c) Subsystem notation projecting the orbitals and number of electrons involved.

using the concepts developed in previous sections and summing all effects. Henceforth, we shall make consistent use of this model unless otherwise stated.

Clearly, we are now one step away from a theory capable of handling any system regardless of complexity. In order to accomplish what we set out to do we must answer the following question: How do we treat "many electrons in many orbitals" systems, or, equivalently, how do we treat many subsystems which are coupled in a one-electron sense?

Consider the transformation of A to B and its description in terms of bond diagrams and associated subsystems. We distinguish two possibilities:

a) The transformation conserves symmetry in a manner which makes possible the definition of elementary subsystems common to A and B. In this case, a detailed theoretical analysis of the problem can be easily carried out. For example, in the transformation shown in Figure 20a, we would conclude that the N $\rightarrow$ N' conversion favors A while the U $\rightarrow$ H' conversion favors B and that the balance of the two effects will be ultimately responsible for the direction of equilibrium.

b) The transformation does not conserve symmetry in a manner which would allow us to define elementary subsystems common to A and B. For example, in the case shown in Figure 20b, we have two options: Either define common elementary subsystems and devise rules for their one electron interaction (Case I) or treat A and B in terms of only one system (Case II). The total system of Case II can become a complex subsystem within some other polyelectronic system.

One of the necessary attributes of qualitative theory is simplicity. With this in mind, we can deal with complex subsystems in an approximate yet sound way. Thus, in the example of Figure 20b, we note that ω_3-$\sigma_2$overlap destruction and ω_1-σ_1 overlap conservation accompanies the A $\rightarrow$ B conversion. This transformation can then be simply viewed as a U to $H^{\ddagger}$ conversion, where the dagger implies that spatial overlap is diminished to some unspecified extent as a result of the A $\rightarrow$ B transformation. Now, we can view a U $\rightarrow$ $H^{\ddagger}$ conversion as having properties intermediate between U $\rightarrow$ H' and U $\rightarrow$ H'' conversions. Similarly, a D $\rightarrow$ $H^{\ddagger}$ conversion can be viewed as having properties intermediate between D $\rightarrow$ H' and D $\rightarrow$ H'' conversions. Finally, a U $\rightarrow$ $D^{\ddagger}$ conversion can be considered as a U $\rightarrow$ D conversion whose energetic benefit is seriously diminished due to loss of overlap. We shall ultimately make use of these ideas in order to formulate a "back of the envelope" theory of chemical bonding.

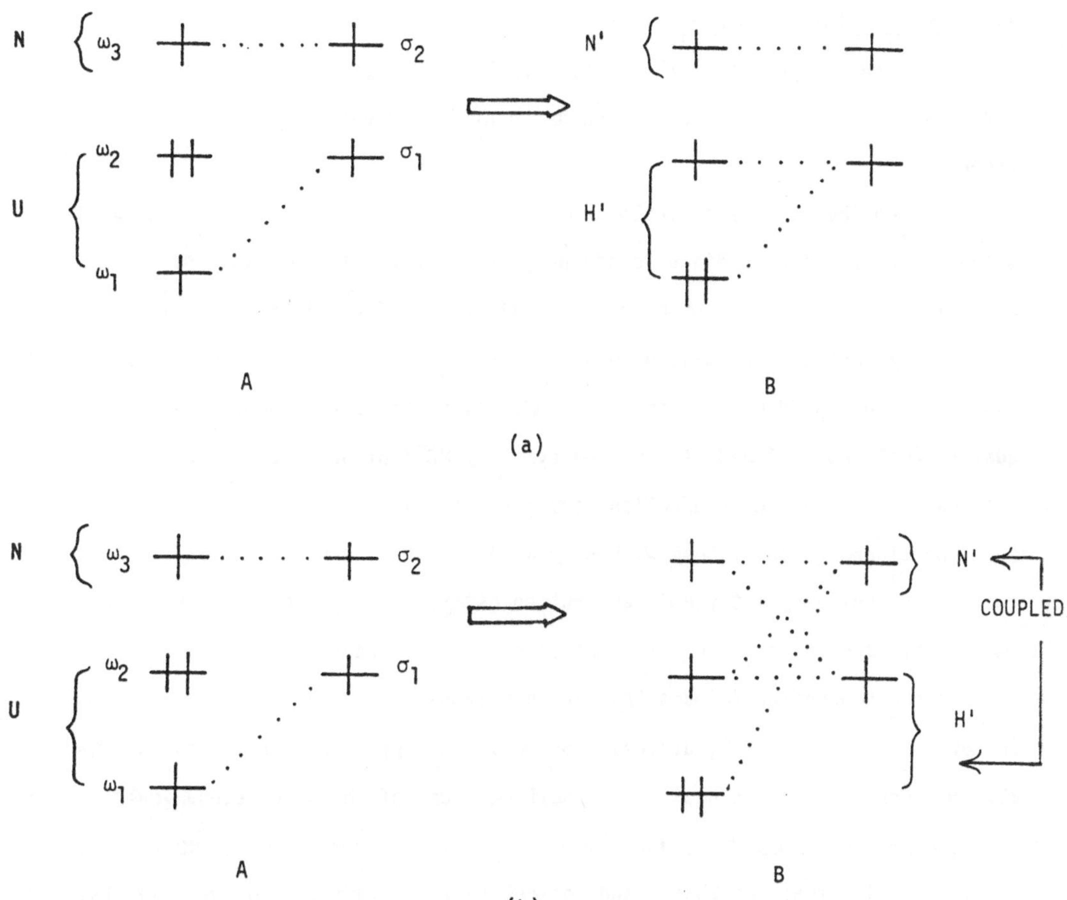

FIGURE 20: a) The transformations of A and B under symmetry constraints which allow the definition of a four electron subsystem over ω_1, ω_2 and σ_1 and a two electron subsystem over ω_3 and σ_2 in both A and B.

b) The transformation of A to B accompanied by removal of symmetry constraints. "Product" B can now be viewed as an N' and an H' subsystem which are coupled in a one-electron sense. Alternatively, it can be viewed as a total H' system where spatial overlap has been impaired.

I. Core and Ligand Group Orbitals

Before we proceed with illustrative applications of MOVB theory some clarifying comments regarding the fundamental ingredients of the theory are in order:

a) When the core is a single atom, the core orbitals are the AO's themselves. In the case of diatomic and polyatomic cores, the orbitals of the central fragment are MO's which can be written down from first principles without any explicit calculation e.g., diatomic core, or they can be computed by some standard procedure. In general, EHMO calculations are adequate for a qualitatively correct ordering of the symmetry MO's of polyatomic core fragments. The same considerations are pertinent to the symmetry orbitals which span the ligands. In systems with only a few ligands, explicit calculations of the corresponding symmetry MO's are not necessary since these can be written immediately from first principles, at least in most cases.

b) The convention for drawing the core symmetry orbitals will be as follows: The MO's of a C_2 diatomic core have the appearance indicated in Figure 21. However, in our discussions we shall make use of the more "economical" notation which emphasizes the principal character of each MO keeping in mind that the bottom and top orbitals ω_1 and ω_8 do not overlap with the ligand orbitals as much as the rest of the core orbitals. Furthermore, we shall continue the practice of symbolizing the core MO's by ω_i and the ligand MO's by σ_i.

c) The proper construction of ligand symmetry orbitals is an essential requirement for a correct application of MOVB theory. The fundamental point which ought to be recognized is that, in general, overlap of nonbonded ligand orbitals is significant, something which guarantees that the ligand symmetry orbitals will be nondegenerate. This implies that a distinction among U, H, and

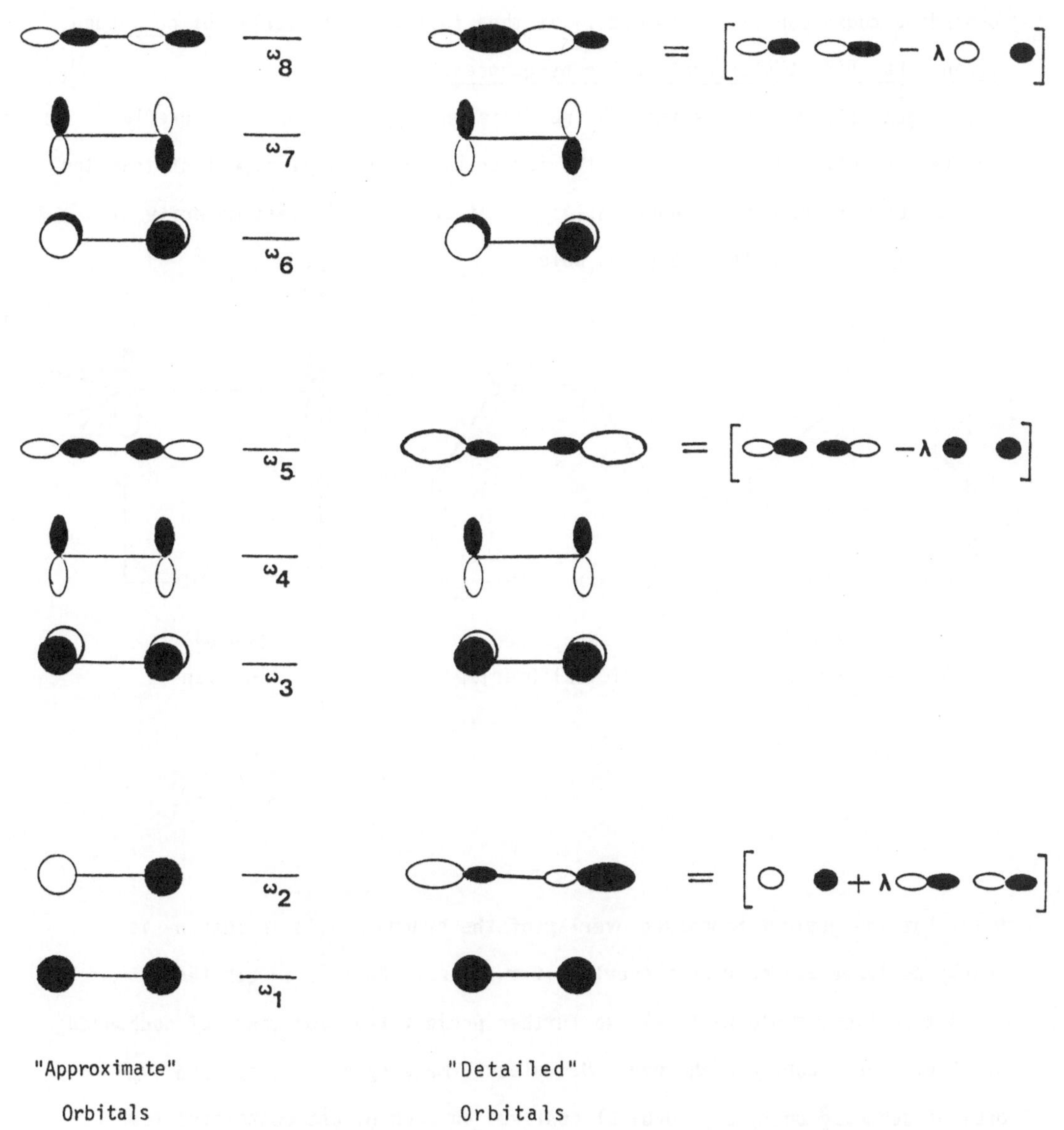

FIGURE 21: The "detailed" and "approximate" forms of the C_2 MO's. By convention, we shall be using the "approximate" forms. Core MO's are symbolized by ω_n.

D bonding modes can always be made. It then follows that overlap of nonbonded ligands has direct stereochemical consequences.

In general, we can distinguish two different types of nonbonded overlap, namely, geminal and vicinal, with the former being much more important than the latter but with both being non-negligible, at least in the vast majority of cases. Typical examples are given below.

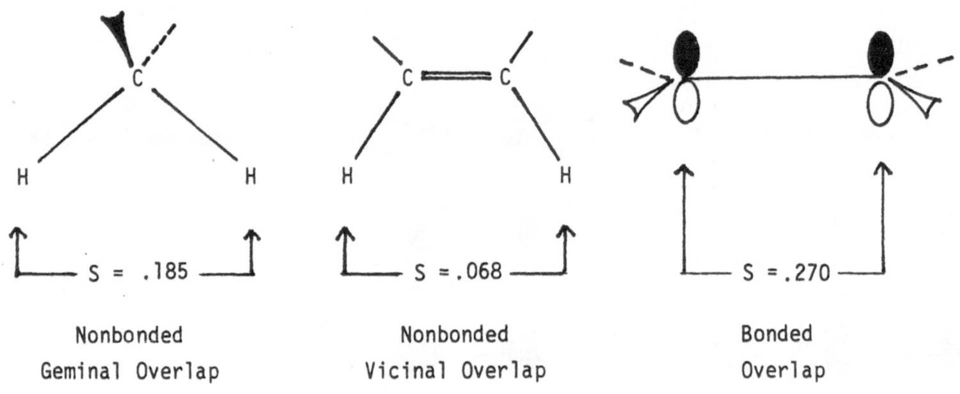

Nonbonded Geminal Overlap

Nonbonded Vicinal Overlap

Bonded Overlap

Note that the geminal nonbonded overlap of the hydrogen AO's in methane is nearly as large as the direct overlap of two carbon 2p AO's in ethylene.

A specific example will help to further project the importance of nonbonded overlap. Thus, consider the bent AH_2 molecule made up from an sp^2 and a p orbital centered on A, an s orbital centered on each H, and containing four electrons. The bond diagram will look like the one shown below. Immediately, the reader will recognize that AH_2 can be viewed as the product of the "forbidden" union of A, with the sp^2 orbital doubly occupied, and H_2.

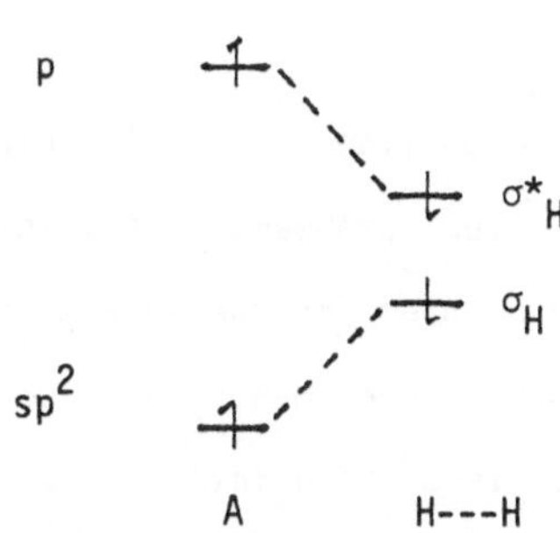

Indeed, nonbonded overlap guarantees that a reference geometry of a molecule can be viewed as the result of a "forbidden" union of core and ligands. In this light, <u>we shall see that the lowest energy geometry of a molecule can be regarded as that which best removes the "forbiddenness" of some other reference geometry</u>.

J. Applications

We are finally prepared to outline a general recipe for the qualitative analysis of ground state structure problems. In a following paper, we shall see that a similar procedure can be used for the treatment of valence shell excited state structural problems. The key elements, in sequence, are as follows:

a) A molecule or complex is divided into a core fragment (C) and a fragment which contains all ligands (L).

b) A bond diagram is constructed for every assumed geometry of the molecule or complex by following these steps:

1. The core and ligand symmetry adapted orbitals are generated either from first principles or by explicit computation.

2. The electrons are arranged in the core and ligand orbitals in a way which generates the lowest energy reference, "perfect pairing", R , CW subject to the symmetry constraints imposed by the geometry in question. This is the lowest energy CW of maximum open shell character which generates maximum overlap attraction between the core and the ligands and wherein the core and ligand fragments have the highest multiplicity. For example, the R CW for a six electron-five orbital system where the two lowest core (ω_1 and ω_2) and the lowest ligand (σ_1) orbitals are of one symmetry type and the highest core (ω_3) and ligand (σ_2) orbitals are of a different symmetry type is written as follows:

	C		L	
ω_3	↑		↓	σ_2
ω_2	↑		↓	σ_1
ω_1	↑↓			

3. The bond diagram for the geometry in question is constructed by adding dashed lines to the drawing of the R CW in order to denote all possible CW's which can be generated under the imposed symmetry constraints. For example, the bond diagram corresponding to the previous case becomes:

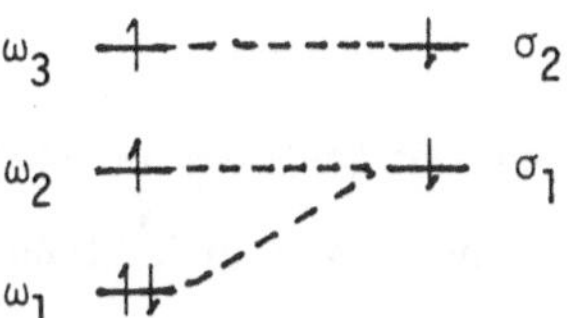

4. In dealing with bond diagrams, we always bear in mind that the "parent" R CW is not necessarily the lowest energy CW, though in many applications this is the case. Pinpointing exactly which CW is the lowest energy one is unnecessary for the prediction of qualitative trends, at least in most cases.

c) The subsystem convention is specified.

d) The result of a structural change on each subsystem is spelled out in the form of an equation. For example, if the structural modification changes an (ω ,σ /2) subsystem from N to N', we write: N(ω ,σ /2) $\rightarrow$ N'(ω , σ /2)

e) The critical subsystem conversion(s) is(are) singled out. This becomes necessary because the energetics of some subsystems may or may not change as a result of a structural modification.

f) A conclusion regarding the effect of the modification, i.e., stabilizing or destabilizing, is reached from appraisal of the critical subsystem conversions according to the following delocalization rules:

1. Hybridization Rule: H bonding becomes increasingly favorable relative to D bonding as primary CT occurs in a direction which prevents overlap repulsion and fosters secondary delocalization.

2. Deexcitation Rule: H bonding becomes increasingly favorable relative to U bonding as primary CT occurs in a direction which turns off secondary delocalization.

3. Competition Rule: Whenever fragment orbitals are well separated in energy, the energy gain accompanying a U to H' transformation or the energy loss accompanying the reverse process dwarfs the energy gain or loss accompanying a D to H' (or vice versa) transformation.

The first two rules can be restated in an alternative language:

1. H bonding becomes increasingly more favorable relative to D bonding as V-type CW's attain low energy.

2. H bonding becomes increasingly favorable relative to U bonding as I-type CW's attain low energy.

The above rules are based on the realization that intrinsic optimization of hybridization or deexcitation is independent of spatial overlap. It is of paramount importance to recognize that these rules are applicable to actual chemical problems only when the structural modification of interest affects the energies of the isolated fragments (F_i or P_i terms) while leaving the interaction energies and especially the overlap interaction (X_i and H_{ij} terms) relatively unchanged. When this condition is not met as a result of nature prohibiting the independent variations of F_i, and X_i, and H_{ij}, exceptions are expected which *in toto* may define a new chemical concept. This problem will be treated in a separate paper.

The following discussion of representative molecules does not aim to shock the reader with previously unfathomable conclusions but rather it attempts to illustrate clearly the "how to do it" aspect of MOVB theory. By contrast, the following papers zero in on problems where there is a fundamental disparity between the conclusions reached on the basis of MOVB theory and those derived on the basis of previous qualitative models.

1. Methane

As a first example of a MOVB theory application, we consider the simple problem of methane isomerism and, in particular, we compare the tetrahedral and (T) and Planar (P) forms of this molecule. The bond diagrams are shown in Figure 22. The subsystem convention is:

$s,\sigma_1/2$	$p_y,\sigma_2/2$	$p_x,\sigma_3/2$	$p_z,\sigma_4/2$

The approximate energetic consequences of the T → P transformation with regards to each subsystem are as follows:

$$N\ (s,\sigma_1'/2) \longrightarrow N\ (s,\sigma_1''/2)$$
$$N\ (p_y,\sigma_2'/2) \longrightarrow N\ (p_y,\sigma_2''/2)$$
$$N\ (p_x,\sigma_3'/2) \longrightarrow N\ (p_x,\sigma_3''/2)$$
$$N\ (p_z,\sigma_4'/2) \longrightarrow N'(p_z,\sigma_4''/2)$$

Clearly, the critical subsystem conversion is the one shown below:

$$N\ (p_z,\sigma_4'/2) \longrightarrow N'\ (p_z,\sigma_4''/2)$$

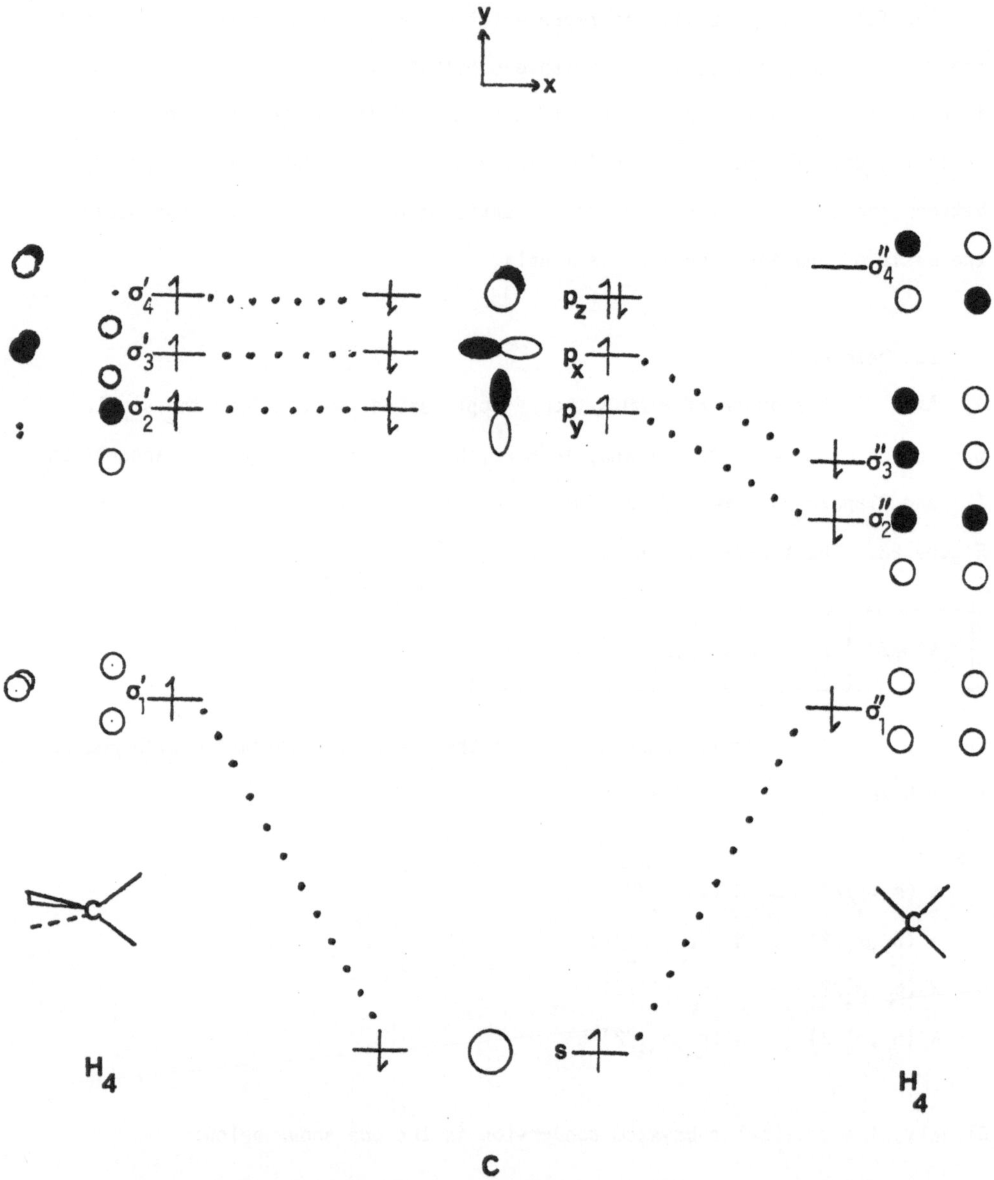

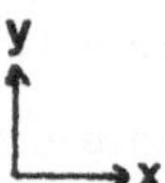

H_4

C

H_4

FIGURE 22: Bond diagrams for tetrahedral (T) and planar (P) methane. Note the four multicenter bonds in T. In P, only three multicenter bonds can be formed. For in-text discussion, tetrahedral ligand orbitals are denoted by σ'_n and planar ligand orbitals by σ''_n.

This amounts to breaking one multicentric bond. Hence, we conclude that T should be heavily favored over P. It is immediately evident that in order to reduce the energy difference between the two forms we must replace the hydrogens by weakly bonding atoms or groups which in addition have low lying vacant orbitals which can be used to restore the fourth bond through overlap with the doubly occupied p_z carbon AO.

The first condition is satisfied by highly electropositive atoms such as atoms of the left two columns of the Periodic Table. This arises because the AO resonance integral , β_{tu}, depends on h_{tt} and h_{uu} which, in turn, depend on the atomic electronegativities [See Equation (37)]. The same atoms also satisfy the second condition by virtue of having np AO's which lie close to the bonding ns AO's. Lithium comes as close as possible to satisfying both of the above conditions. Schleyer and coworkers have studied the T-P energy gap in methane and its derivatives and they have found that Li shrinks and may even reverse the relative energetic order of the T and P forms.[32] The interpretation of the structures of lithiated hydrocarbons will be presented in a following paper.

2. BeH_2

The simple triatomic BeH_2 provides us with the first example of a conflict between spatial overlap maximization and deexcitation. The bond diagrams for Linear (L) and Bent (B) forms are shown in Figure 23. Neglecting the vacant p_z AO, the subsystem convention is:

$s,\sigma_1,p_y/2$	$p_x,\sigma_2/2$

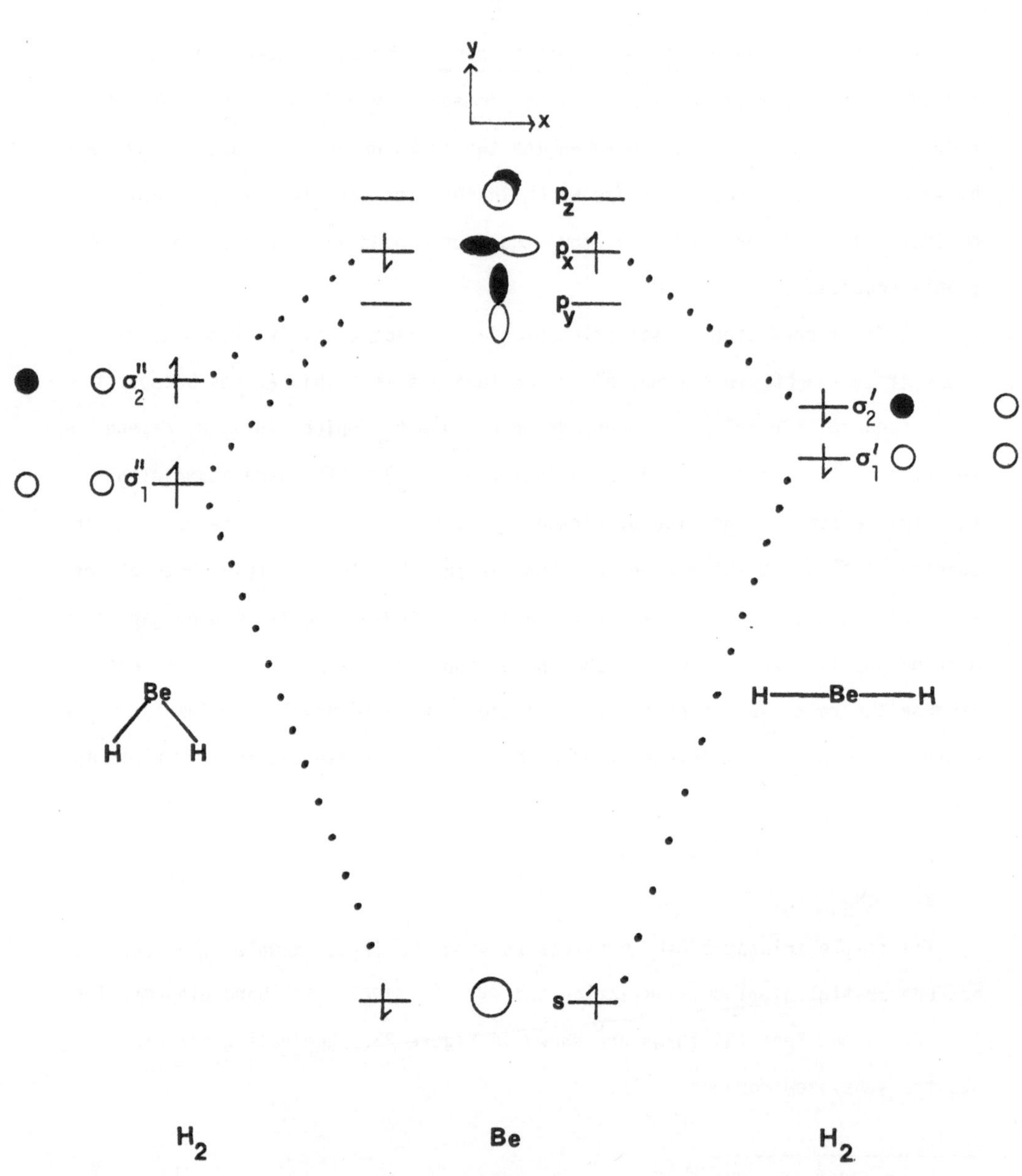

FIGURE 23: Bond diagrams for bent (B) and linear (L) BeH_2. Note how primary CT in the H' subsystem of the B form cannot promote secondary CT, i.e., the B form is ineffectively hybridized. For in-text discussion, bent ligand orbitals are denoted by σ_n'' and linear ligand orbitals by σ_n'.

The energetic consequences of the L$\rightarrow$ B transformation are as follows:

$$N\ (p_x, \sigma_2'/2) \longrightarrow N'\ (p_x, \sigma''_2/2)$$
$$D\ (s, \sigma_1', p_y/2) \longrightarrow H'\ (s, \sigma''_1, p_y/2)$$

Clearly, both subsystem conversions are critical. Specifically, the two electron-two orbital subsystem defines a multicentric bond which is stronger in L because of greater spatial orbital overlap. Furthermore, the two electron-three orbital subsystem constitutes a D-type bond in the L geometry and an H'-type bond in the B geometry. Since we already know that H' is always more favorable than D in two electron-n orbital systems, bonding in this subsystem now favors B over L. In BeH_2 the conflict is resolved in favor of the spatial overlap factor and BeH_2 ends up being linear. Now, the importance of this example is considerable because it typifies the essential conflict we shall encounter in stereochemical problems time and again. Specifically, U $\rightarrow$ H' or D $\rightarrow$ H' bond transformation is accompanied by concurrent N $\rightarrow$ N' transformation. In this case, our qualitative understanding of the problem does not allow us to make a secure absolute prediction since a quantitative evaluation of two opposing factors is needed. However, it is clear that the energy gain which accompanies the D $\rightarrow$ H' transformation in BeH_2 is small for two reasons:

a) The energy gain in a D $\rightarrow$ H' conversion is intrinsically small because extra delocalization in H relative to D can occur only by using higher lying orbitals, i.e., via core excitation.

b) V makes a small contribution to the wavefunction of the subsystem as σ_1 lies much above the 2s AO of Be, or, equivalently, primary CT does not lead to secondary CT.

Hence, we anticipate that the L form will be preferred over the B form.

While we could not make a safe a priori prediction of the geometry of BeH_2, we can design related systems where the tendency for bending will be maximized. This can be done in two different ways:

a) By tuning the D $\rightarrow$ H' conversion.

b) By tuning the N $\rightarrow$ N' conversion.

We restrict our attention to the first possibility. In this case, enhanced bending tendency can be brought about by a structural modification which lowers the energy of V relative to the rest of the CW's. This is best achieved by replacing the hydrogens by more electronegative ligands and Be by a more electropositive core, i.e., by moving towards the right along a period of the Periodic Table insofar as univalent ligands are concerned and down along a column of the Periodic Table insofar as divalent central atoms are concerned.

The way in which d-orbitals of the central atom can influence the geometry of a triatomic molecule like BeH_2 is intriguing. The bond diagrams of Figure 24 clearly illustrate that the d_{xy} "hole" plays exactly the same role with respect to the p_x-σ_2 two electron multicenter bond that the p_y "hole" plays with respect to the s-σ_1 two electron multicenter bond. Specifically, the transformation of L to B is now described by the following two subsystem transformations.

$$D\ (s, p_y, \sigma_1'/2) \rightarrow H'\ (s, p_y, \sigma_1''/2)$$
$$D\ (p_x, d_{xy}, \sigma_2'/2) \rightarrow H''\ (p_x, d_{xy}, \sigma_2''/2)$$

Both subsystem transformations will become more favorable as CT occurs from core to ligands, i.e., under the conditions described above. Thus, d-orbitals are expected to reinforce a nascent tendency for bending already present in their absence.

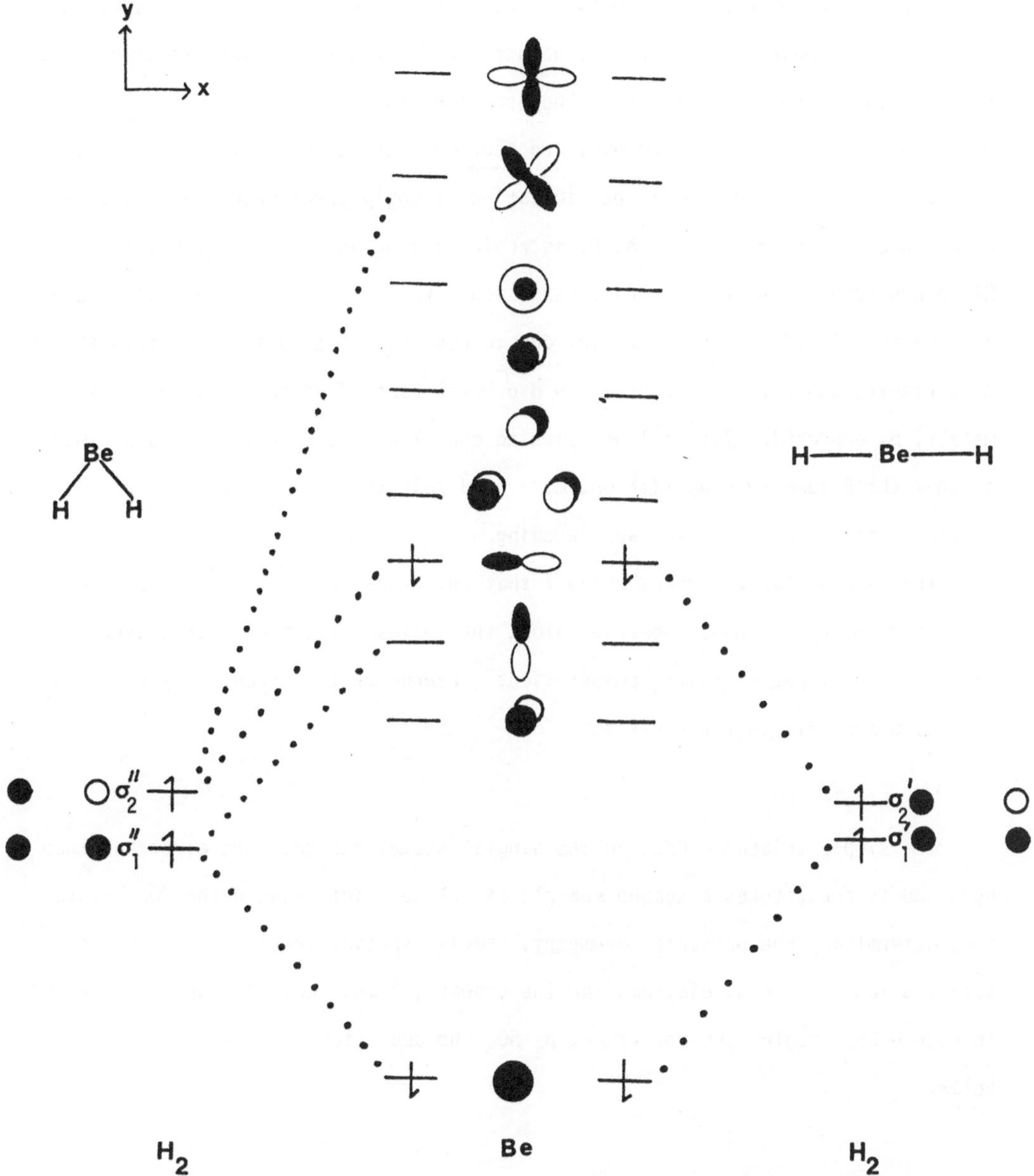

FIGURE 24: The additional effect of d-orbital participation on the relative energies of bent (B) and linear (L) BeH_2. Note that, of the s-σ_1 and p_x - σ_2 bonds, only the latter can promote strong secondary CT to d-orbitals in the bent form. The d_z2 and $d_{x^2-y^2}$ orbitals are not considered because they do not affect the argument as they are as additional "holes" with respect to the s - σ_1 multicenter bond.

Replacement of H by F not only redirects interfragmental CT and creates a tendency for bending in BeF_2 but also introduces the all important fluorine lone pairs which can combine with the d "holes" of Be in order to define three multicenter bonds in the linear geometry and four multicenter bonds in the bent geometry. As a result, d orbital participation strongly predisposes BeF_2 towards adoption of a bent geometry. As Be is replaced sequentially by Mg, Ca, Sr, and Ba, bonds become longer, nonbonded repulsion (overlap and "classical")is reduced, and the associated bending restraint due to nonbonded interaction is diminished. As a result, the influence exerted by d orbital participation increases. Ultimately, by converting ligand lone pairs to core-ligand bond pairs in such a way so that the B form ends up with one more bond pair than the L form, d orbital participation is expected to cause bending

The data of Table 3 makes evident that the predicted switchover from linear to bent geometry is indeed observed along the series BeF_2, MgF_2, CaF_2, SrF_2, and BaF_2. Furthermore, ab initio computations[33] reproduce this trend only by inclusion of d orbitals in the basis set.

3. Singlet CH_2

The simple triatomic CH_2, in its singlet state, has two more electrons than BeH_2 and it constitutes a second example of the conflict between the two key factors determining the preferred geometry, namely, spatial overlap maximization and deexcitation. The bond diagrams for the Linear (L) and Bent (B) forms are snown in Figure 25. Neglecting the vacant p_z AO, the subsystem convention is shown below:

$s,\sigma_1,p_y/4$	$p_x,\sigma_2/2$

Table 3. Bond Angles of AF_2 Molecules.

Molecule	Bond Angle	Reference
BeF_2	180°	a
MgF_2	160°	b
CaF_2	140°	c
SrF_2	108°	c
BaF_2	100°	c

a. Warton, L.; Berg, R.; Klemperer, W. J. Chem Phys. 1974, 40, 3471.

b. Akishin, A.; Spiridonov, V. P. Kristallografiya 1957, 2, 475.

c. Calder, V.; Mann, D. E.; Seshardi, K. S.; Allavena, M.; White, D. J. Chem. Phys. 1969, 51, 2093.

The energetic consequences of the L → B transformation are as follows:

$$U\ (s, \sigma_1', p_y/4) \rightarrow H'\ (s, \sigma_1'', p_y/4)$$
$$N\ (p_x, \sigma_2'/2) \rightarrow N'\ (p_x, \sigma_2''/2)$$

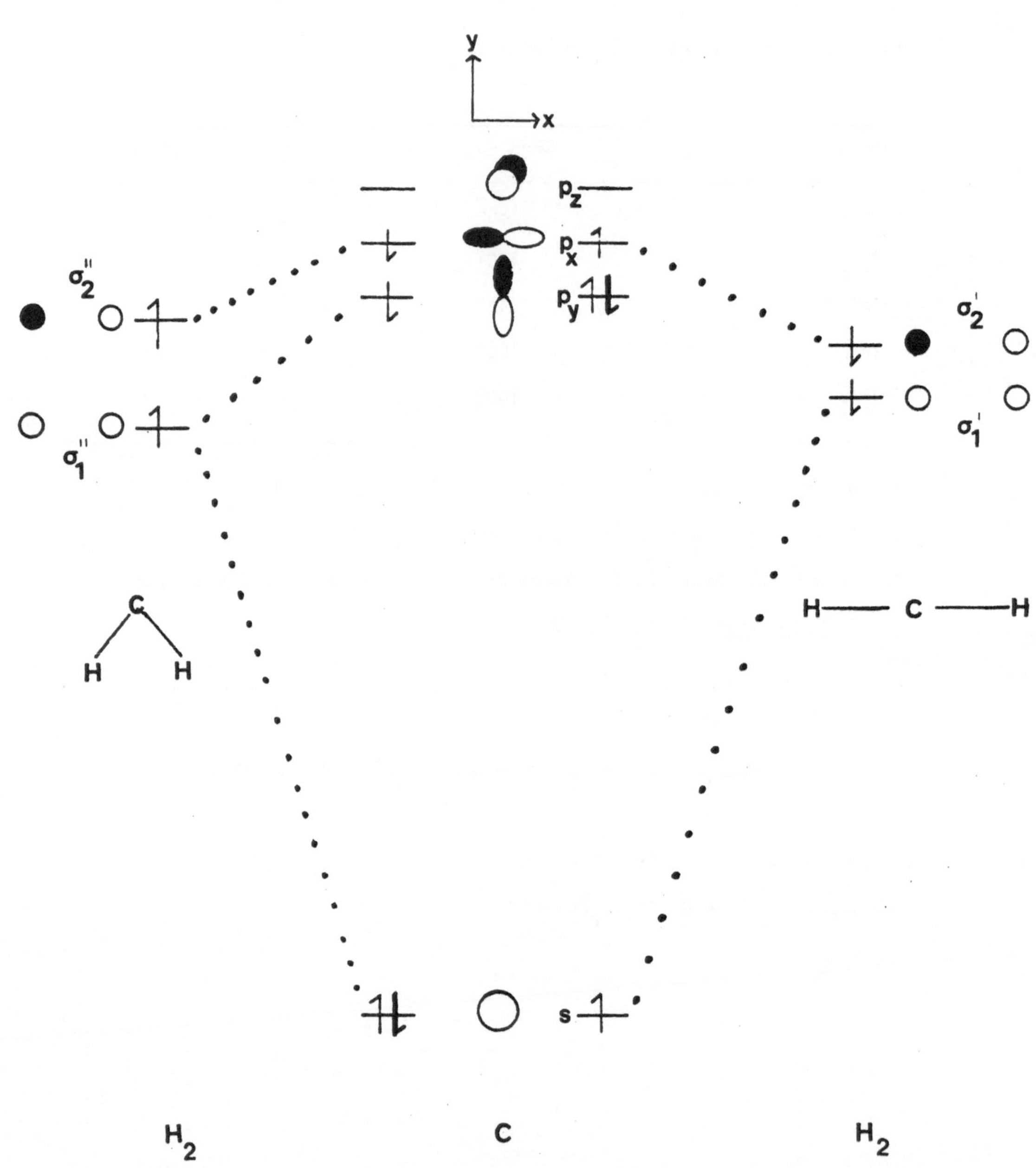

FIGURE 25: Bond diagrams for bent (B) and linear (L) singlet CH_2. Note the deexcitation of the electron in p_y upon bending. For in-text discussion, bent and linear ligand orbitals are denoted by σ_n'' and σ_n', respectively.

Clearly, the critical subsystem conversions are both of the above. The following similarities and differences with respect to the previous BeH_2 problem are apparent:

a) The two electron-two orbital subsystem defines a multicentric bond which is stronger in L because of greater spatial overlap, much like in BeH_2.

b) The four electron-three orbital subsystem constitutes a U bond type in the L geometry and an H' bond type in the B geometry. Since H' is always more favorable than U in four electron-n orbital systems, delocalization in this subsystem favors B over L. The critical difference between BeH_2 and CH_2 is the differing hybridization tendency. In the former case, we deal with an intrinsically small energy gain accompanying the D → H' transformation. By contrast, in CH_2 we deal with a much larger energy gain accompanying the U → H' transformation. The reasons behind the different hybridization tendencies represented by the U → H' and D → H' conversions have been discussed before. In particular, a U → H' transformation in CH_2 represents deexcitation while the D → H' transformation in BeH_2 represents merely extra delocalization to higher lying orbitals and the former conversion is much more beneficial, in an energetic sense, than the latter, as test computations vividly demonstrate (see Figures 14 and 16).

We conclude that the tendency for bending will be much greater in CH_2 than in BeH_2 and that it will probably dominate the adverse effect of loss of spatial overlap in the $(p_x, \sigma_2/2)$ subsystem. The difference of the two systems is the exclusive result of the different magnitude of energy gain which attends D → H' (in BeH_2) and U → H' (in CH_2) conversions. To our knowledge, this constitutes the first physical explanation of the different shape of triatomics having valence shells which differ by one electron pair. It is also a first application of the Competition rule enunciated above. The lowest energy singlet state of CH_2 is known to be bent.[34]

4. C_2H_2

In a previous section we started with an understanding of linear BeH_2 and sought to design rationally a related bent molecule. We achieved this by enhancing the energy gain accompanying the D → H' transformation via substitution of Be and H_2 by a more electropositive core and more electronegative ligands, respectively. We now start from bent CH_2 and seek to design rationally a related linear molecule by diminishing the energy gain accompanying the U → H' transformation. The latter energy gain is due primarily to deexcitation of one electron across a large energy gap separating the 2s and 2p AO's of carbon. Accordingly, we must seek to design a molecule with a much smaller energy gap so that the driving force accompanying the U → H' transformation will be diminished to the extent that the deexcitation factor will now be dominated by the spatial overlap factor favoring the linear form. The bond diagrams of Figure 26 make clear that Linear (L) and Trans (T) acetylene are the simplest representatives of a large number of unsaturated compounds which meet the specifications of our design plan. Indeed, a comparison of the bond diagrams for CH_2 and C_2H_2 makes clear the following:

a) The sigma bonding of L CH_2 is identical in type to that in L C_2H_2.

b) The sigma bonding of B CH_2 is identical in type to that in T C_2H_2.

c) The principal difference between the two molecules, insofar as sigma bonding is concerned, lies in the different magnitude of the energy gap across which deexcitation of an electron must occur. Specifically, the 2s-2p energy gap in C (in CH_2) is almost twice as large as the ω_2-ω_4 energy gap in C_2 (in C_2H_2) for reasons which are trivially simple to understand.[35] Accordingly, we expect that C_2H_2 will have an excellent chance of being linear as is actually found to be the case.

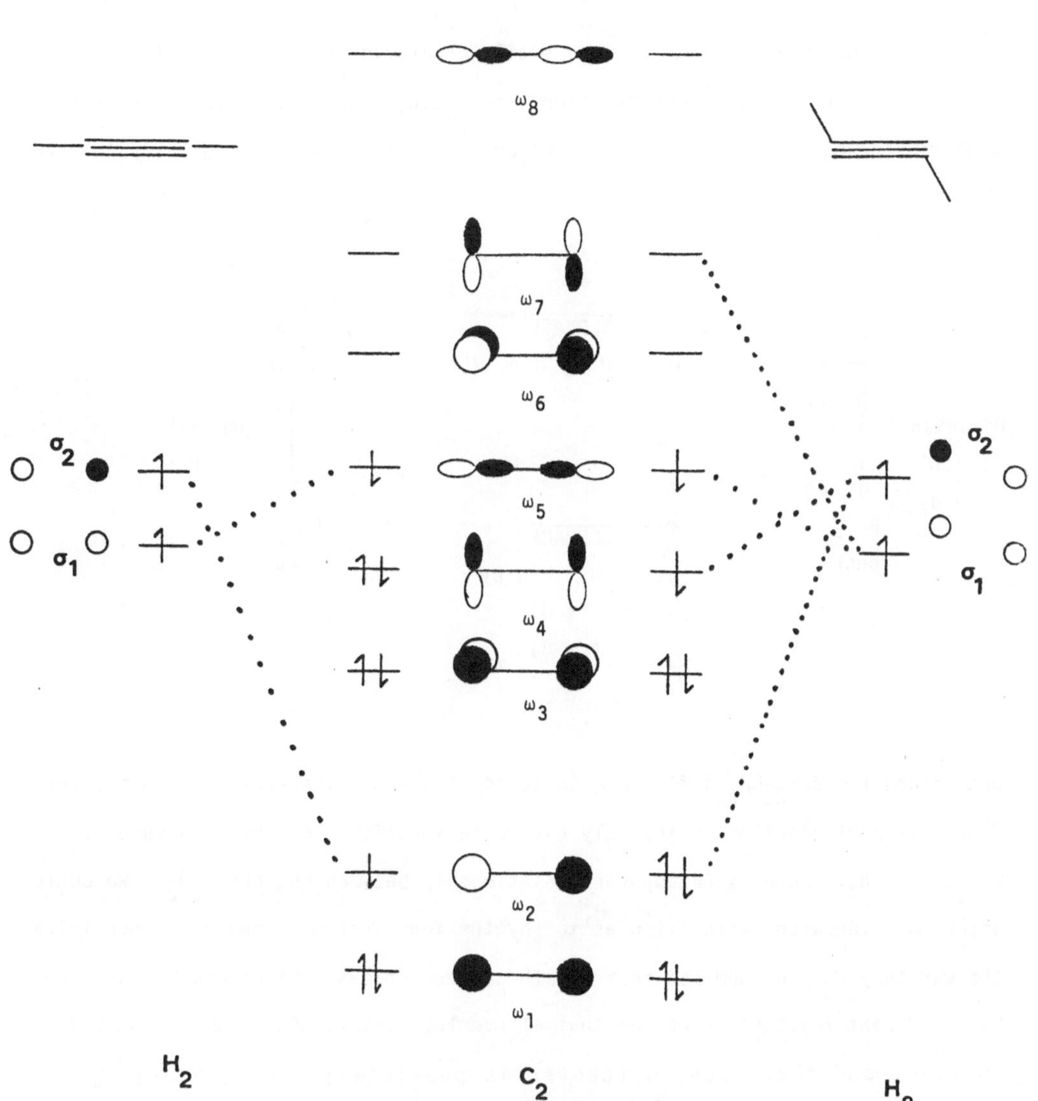

FIGURE 26: Bond diagrams for trans (T) and linear (L) acetylene. For in-text discussion, core orbitals are denoted by ω_n. The role of ω_1 and ω_8 is neglected.

The microcosm represented by the small polyatomics of these last three sections and the required optimizations for making a transition from linear to bent geometries is shown graphically below. Some years ago we would say that we

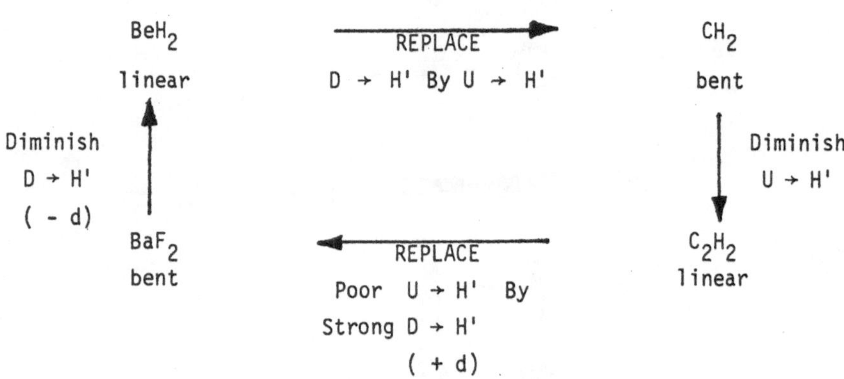

understood the BeH_2-CH_2 difference in terms of the Mulliken-Walsh approach, that d orbital participation is the only way to rationalize the bent structure of BaF_2, and that there is no apparent relationship between CH_2 and C_2H_2. We could offer no integrated explanation as to why the four systems shown above hybridize the way they do. We now see that all these geometrical switchovers are nothing but different resolutions of the spatial overlap maximization - deexcitation conflict and that each switchover is predictable and subject to theoretical design.

5. "Conflicts of Interest" in Molecular Stereochemistry

The application of MOVB theory to BeH_2, CH_2, and C_2H_2 revealed that there is a class of molecules which can exist in isomeric forms A and A' so that A is favored by spatial overlap and A' by deexcitation. In general, A can be converted to A' by means of angular deformation or bond rotation. Henceforth, we shall refer to A and A' as ambivalent isomers or forms in order to project the fact that in transforming A to A' some bonds are weakened while the excitation energy diminishes so that the the substrate is "ambivalent" as to which of the two geometries to adopt. Structural modification of the substrate can direct it towards adoption of one or the other form in a way which is fully predictable on the basis of the MOVB theory.

The idea of ambivalent isomers is not familiar to chemists, for two main reasons:

a) The theory of chemical bonding espoused in the pioneer work of Pauling was based on an approximate form of VB theory and the concepts were developed from consideration of highly simplified models such as two electron-two orbital systems. As a result, optimal hybridization could not be predicted *a priori* but it was invoked *a posteriori*. For example, we are taught that carbon is nearly unhybridized in CH_2, sp hybridized in acetylene, etc., without providing any explanation for why carbon "decided" to adopt the hybridization it did.

b) MO theoretical analyses have further misdirected the thinking of chemists by placing undue emphasis on the implications of overlap populations. In turn, this type of approach has created the impression that one isomer is more stable than another because the bonds of the former are stronger than those of the latter. Clearly, the bonds of linear CH_2 are stronger than those of bent CH_2 yet it is the latter form which is more stable. In other words, the correct thing to say is that A is more stable than A' if it has lower energy as a whole and not necessarily because it has stronger bonds. By eliminating these incorrect preconceptions, not only do we begin to understand the origin of stereochemical diversity in nature but we also become able to design "unconventional" systems in a rational rather than haphazard way. This will be illustrated clearly in a subsequent paper dealing with the structure of lithio derivatives of hydrocarbons.

In summary, we have come to recognize the key factors which determine stereoselection, in the most general sense of the word. The first can be termed the spatial overlap factor and it refers to the tendency of a molecule or complex to adopt a geometry which maximizes spatial overlap. The second can be termed the excitation factor and it refers to the tendency of a molecule to adopt a geometry which allows electron pairs to occupy the lowest energy orbitals available. This important realization has been made possible by the physically meaningful formalism of MOVB theory with Core-Ligand dissection. It now opens the way towards the rational design of unusual molecules, complexes, reactions, etc., via the proper manipulation of the spatial overlap and excitation factors.

6. Cis versus Trans Acetylene

In qualitative theory, a mathematical proof of the electronic origin of a given chemical phenomenon cannot be given because of the assumptions one must necessarily employ. Rather, it is the nature of the argument which makes it believable or reasonable. For example, we are willing to accept the explanation that 1,3-butadiene closes thermally to cyclobutene via a conrotatory rather than a disrotatory motion because it is explicity obvious that the two processes involve transition state complexes which belong to different point groups, in the sense that an axis of symmetry characterizes conrotation and a plane of symmetry characterizes disrotation. As a result, the different symmetries of the MO's of reactants and products become ultimately responsible for the energetic preference of con-over dis-rotation. The same theoretical considerations can be extended to the problem of geometrical isomerism of C_2H_2, where the Trans (T) form has an axis of symmetry and the Cis (C) form a plane of symmetry.

The bond diagrams for C and T acetylene are shown in Figure 27. The simplicity of this problem is apparent. The only difference between the two forms is the nature of hybridization of the two subsystems, with spatial overlap remaining constant. It is clear that the T form is energetically superior to the C form. The physical description of the energetic preference for trans bending is as follows.

a) In the T form, the electron pair occupying the ω_5 and σ''_1 orbitals is shifted towards the σ''_1 orbital without "feeling" any overlap repulsion by the bonding electron pair of ω_4. In this way, subsequent delocalization to the vacant orbital ω_7 is possible. By contrast, in the C form, the same electron pair is shifted toward σ'_1 in a way which "turns on" overlap repulsion with the doubly occupied orbital ω_4 and "turns off" delocalization into the vacant orbital ω_7.

b) In the T form, the electron pair occupying the ω_2 and σ''_2 orbitals is shifted down towards the ω_2 orbital, thus creating a vacancy into which the bonding pair of ω_4 can now delocalize. By contrast, in the C form, the same electron pair is shifted towards orbital ω_2 in a way which prevents effective delocalization into the vacant orbital ω_7.

The descriptive statements above can be replaced by the adorned bonding diagram of Figure 28 which emphasizes the direction of primary CT which renders T energetically superior to C. Such representation makes the key features of the problem self evident and obviates the need for lengthy discussion.

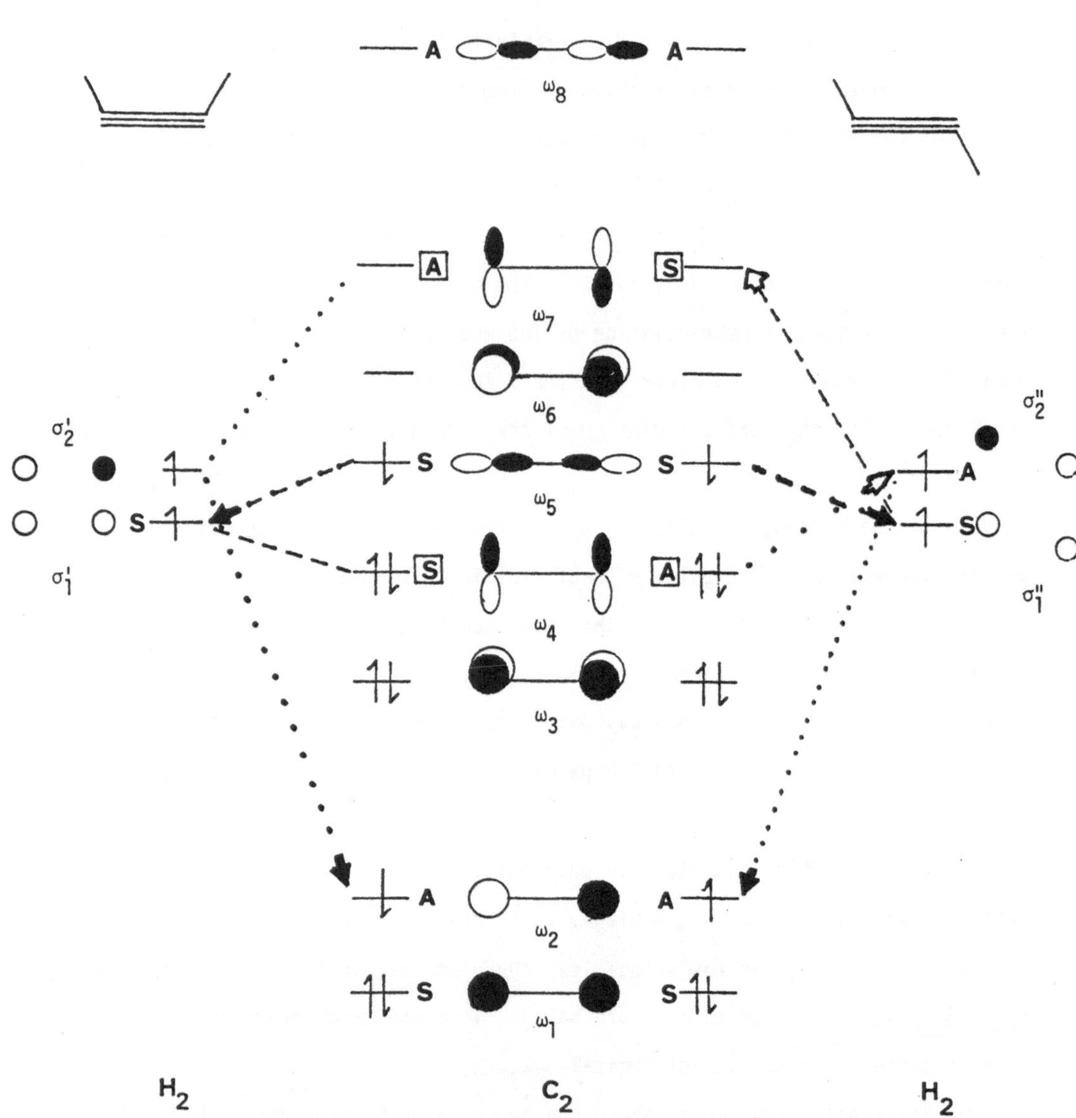

FIGURE 27: Bond diagrams for cis and trans acetylenes showing how directional CT favors the trans isomer. Solid arrowheads denote primary CT and open arrowheads denote secondary CT. For in-text discussion, the trans and cis ligand orbitals are denoted by σ_n'' and σ_n', respectively, and the core orbitals by ω_n. The The roles of ω_1 and ω_8 are neglected.

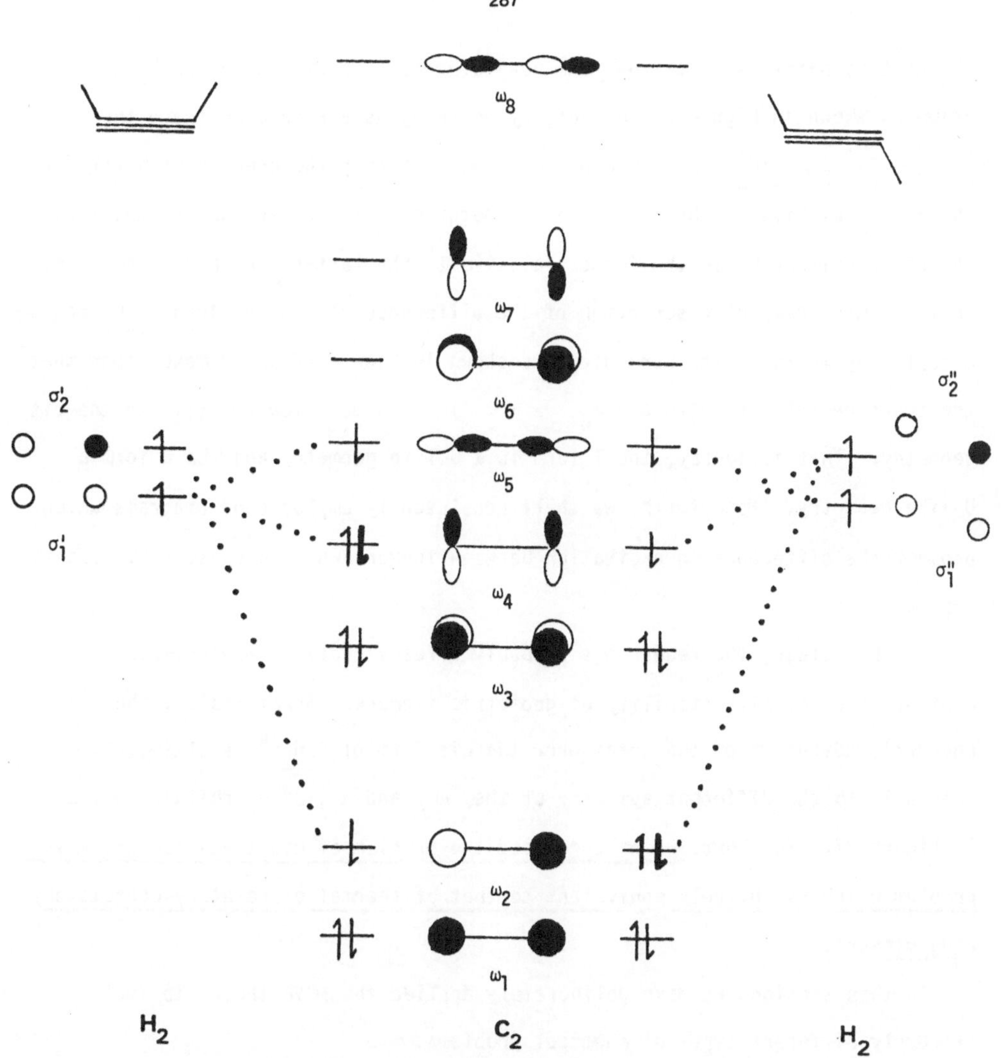

FIGURE 28: Bond diagrams for trans (T) and cis (C) acetylene. For in-text discussion, trans and cis ligand orbitals are denoted by σ''_n and σ'_n, respectively, and core orbitals by ω_n.

At this point, we note that the bond diagrams for cis and trans C_2H_2 were drawn as shown in Figure 27 so that, by defining as a reference frame the CW $\omega_1^2 - \omega_2^1 - \omega_3^2 - \omega_4^2 - \omega_5^1 - \sigma_1^1 - \sigma_2^1$, we could project the fact that the greater stability of the trans relative to the cis isomer is nothing but a consequence of the more favorable primary CT in the former relative to the latter. An equivalent and, in fact, more compact description of the difference of cis and trans C_2H_2 can be effected by means of the bond diagrams shown in Figure 28 which make clear that the trans geometry permits an $\omega_4 \rightarrow \omega_2$ electron demotion relative to the cis geometry. That is to say, the T form is a D-like geometry and the C form a U-like geometry. Henceforth, we shall consistently employ bond diagrams which project the difference in excitation between isomers wherein spatial overlap is kept constant.

At this stage, the reader has probably already realized how symmetry controls the relative stability of geometric isomers. Specifically, the energetic advantage of the trans over the cis form of C_2H_2[36] is ultimately traceable to the different symmetry of the ω_4 and ω_7 core orbitals indicated in Figure 27. In short, we have reduced the problem of cis-trans isomerism to a problem which is entirely equivalent to that of thermal disrotatory-conrotatory ring closure.

In this section, we have deliberately applied the MOVB theory to two extremely different types of chemical problems:

a) "Controversial" problems, where a stereochemical transformation is accompanied by a "bonding conflict" which is resolved in favor of one or the other isomer depending upon the identity of the component fragments.

b) "Noncontroversial" problems where spatial overlap is kept constant and orbital symmetry alone becomes the sole determinant of stereoselection.

Theoretical treatments of problems of the first type are often derisively termed "fuzzy" or "unclear" exactly because they bring into focus the realities of conflict at the electronic level, thus, denying a simplistic view of chemistry. On the other hand, theoretical treatments of problems of the second type are often called elegant because of the unidirectionality of effects. Our viewpoint is somewhat different: Systems of the noncontroversial type are interesting and important. However, it is exactly the lack of conflict which withholds from us the possibility of manipulating them in order to produce chemically unexpected results. By contrast, the "controversial" systems are frought with exciting possibilities. Once the key conflicts at the electronic level are clearly comprehended, we can make creative use of theory to design novel compounds and discover new mechanisms.

K. Compact MOVB Theory

The analysis of the various stereochemical problems presented in the previous section has been a detailed one. Oftentimes a careful consideration of subsystems can assist us in developing a thorough understanding of the energetic consequences of a stereochemical change. In this work we have used the detailed approach not only for this reason but also with the purpose of exemplifying adequately a new way of thinking about molecules and reactions. However, in applied theory, simplicity is a virtue of paramount importance. How can we render the analyses presented above "back of the envelope" analyses? The answer is very simple: Instead of enumerating all subsystems and the energetic consequences of stereochemical change for each subsystem, we can view the entire system as a U, H, or D type system. Furthermore, placement of a dagger next to the bond type symbol can be taken to denote the form in which spatial overlap has been impaired. For example, the transformation of L to B BeH_2 can be viewed as a D to $H^{\ddagger}$ conversion.

The question now arises: Can the dagger be associated with any of D, H, or U or, rather, with only one of the three intrinsic bonding forms? It is not difficult to see that the requirement, by definition, of more extensive orbital mixing in H relative to D or U is satisfied if the ligands are displaced away from the positions of maximum orbital overlap, or, equivalently, if D and U have higher symmetry than H. As a result, the dagger is expected to be associated with H, at least in the majority of applications. Typical examples of how symmetry reduction leads to spatial overlap reduction and hybridization are shown below:

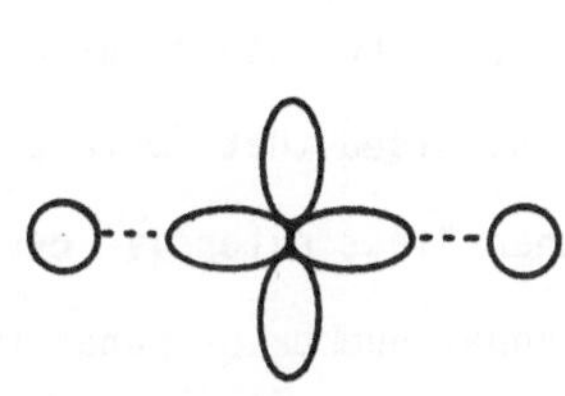

LINEAR AH_2

STRONG p_x - s OVERLAP
NO HYBRIDIZATION

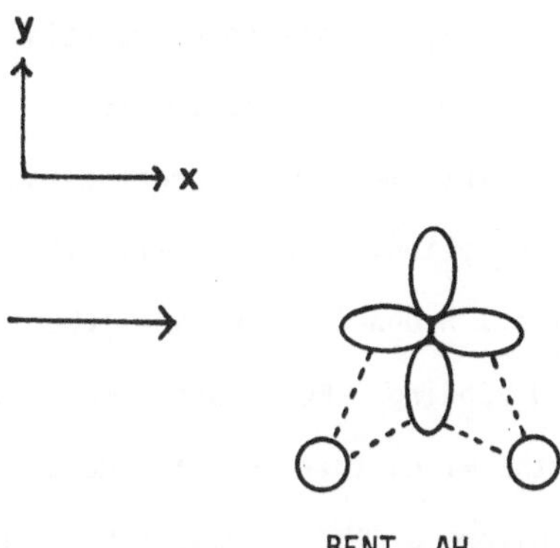

BENT AH_2

WEAK p_x - s AND p_y - s OVERLAP
HYBRIDIZATION

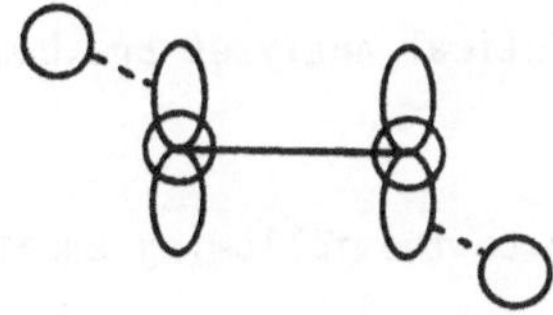

TRANS A_2H_2

STRONG p_y - s OVERLAP
NO HYBRIDIZATION

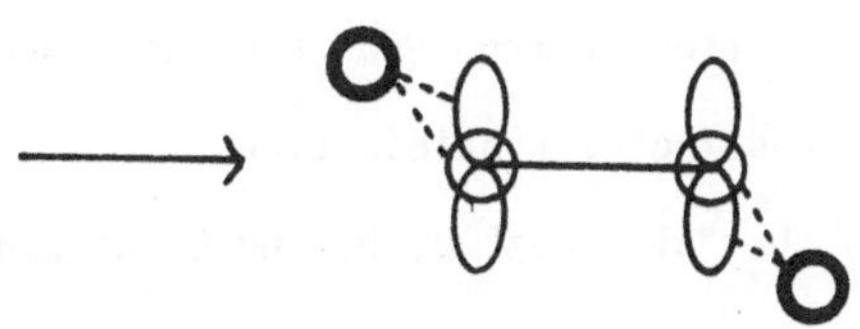

GAUCHE A_2H_2

WEAK p_y - s and p_z - s OVERLAP
HYBRIDIZATION

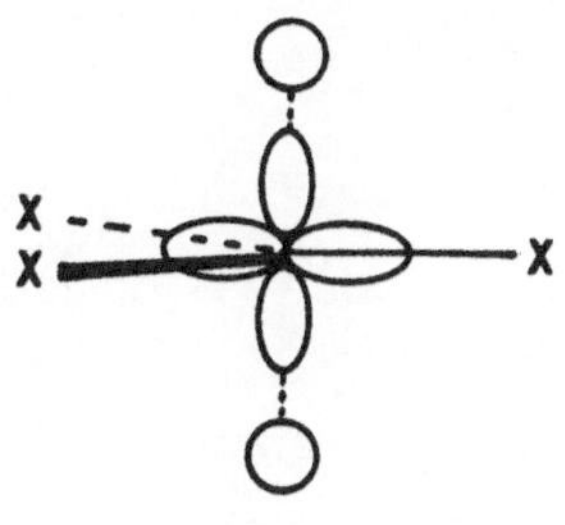

AX_3H_2

STRONG p_y - s OVERLAP
NO HYBRIDIZATION

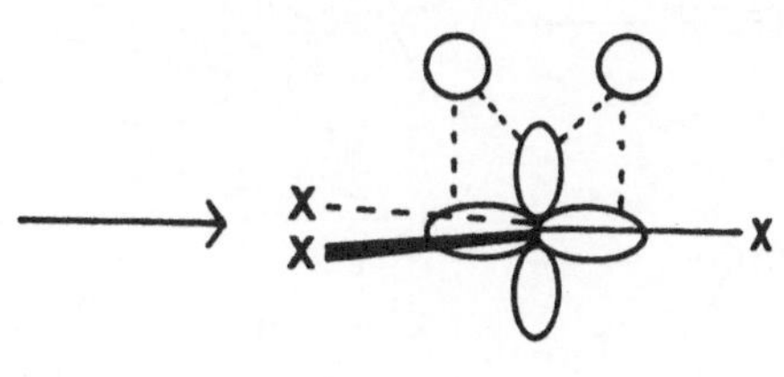

AX_3H_2

WEAK p_y - s and p_x - s OVERLAP
HYBRIDIZATION

In the above examples, the core orbitals which maintain constant overlap with the ligand orbitals are omitted.

At this point, we open a parenthesis in order to bring into focus a very common misconception. Specifically, it is widely surmised that "more extensive overlap", i.e., H bonding, implies lower energy than "less extensive overlap", i.e., U or D bonding. For systems made up of an equal number of bonds (multi-center or otherwise) this is not true at all. Optimal bonding depends on the electronic nature of the component parts of the system under scrutiny in a way which is qualitatively predictable by the selection rules of Section J . The misconception we spoke of above arises because MO theory obscures the fact that <u>more extensive orbital mixing does not mean more bonds</u>. It merely means H bonds which are different from U or D bonds. Many theoretical analyses are based on a misunderstanding of hybridization.

With the "alphabet" U, H, and D, we can now write the following equations:

H-Be-H ⟶ bent H–Be–H

D → $H^{\ddagger}$

H-C̈-H ⟶ bent H–C̈–H

U → $H^{\ddagger}$

H-C≡C-H ⟶ trans-bent H–C≡C–H

U → $H^{\ddagger}$

cis-bent H–C≡C–H ⟶ trans-bent H–C≡C–H

U → D

It is now evident that in comparing two different geometries of a given system and seeking ways to manipulate their relative energies one must perform the following four simple tasks:

a) Write down the corresponding bond diagrams.

b) Associate the appropriate label U, H, or D (with or without a dagger superscript) with each bond diagram. In order to do so, one needs to ascertain relative excitation energies and the different extent of fragment orbital overlap from examination of the R CW, i.e. the "perfect pairing" CW implied by the bond diagrams as written.

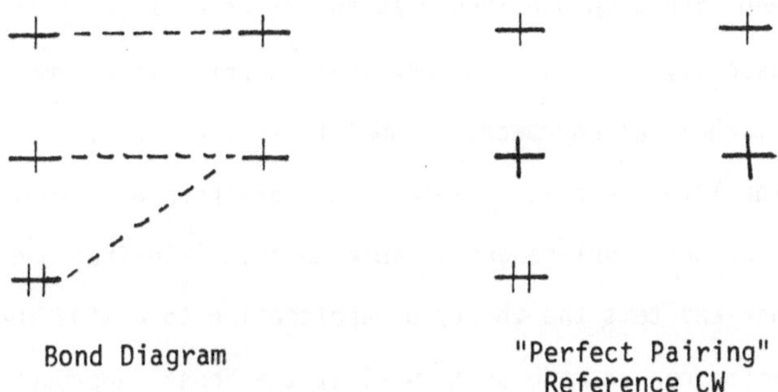

Bond Diagram "Perfect Pairing" Reference CW

c) Predict the relative energies of the two geometries. This can be done unambiguously if both geometries carry the same label. If the two systems carry different labels, their relative energies become a function of the electronic nature of the components of the system, i.e., the core and the ligands. We say that the system is subject to design.

d) Design is done by reference to the delocalization rules of Section J.

With U, H, and D now forming the alphabet of compact MOVB theory, we expect to encounter an entire gamut of possibilities. Thus, depending on the magnitude of excitation energy, we anticipate a switch of the relative stability order U>H to H>U. Similarly, depending on the direction of primary CT and the spacing of the energy levels of core and ligands, we expect a switch of the relative stability orders H>D to D>H. The only invariant stability order is D>U. The rational procedure for effecting the switchovers is embodied in the various rules presented in Section J. In short, we can now begin to understand the electronic basis of stereochemical diversity.

Let us now digress for a moment in order to consider the predicament of the experimentalist who wishes to use theory as an interpretative and/or predictive tool without necessarily becoming a theoretical expert. In attempting to select the "best" quantum chemical approach, a chemist of such predilection soon arrives at an impasse: The literature is replete with a plethora of formulisms and accompaying applications. Unless one is able to translate from one theoretical language to another and test the theory by application to a wide spectrum of problems, no decision can be made as to what is the "best" approach. Ultimately, this leads to a proliferation of implicitly related models and aborts a self consistent, non-illusory understanding of chemistry. In this context, it is not injudicious to reiterate precisely what is "new" about the qualitative MOVB theory of chemical bonding presented in this work. Thus, if we arbitrarily define Single Determinant (SD) MO theory as the frame of reference, we can make the following statement: *MOVB theory represents a conceptual and formal improvement over SD MO theory. Furthermore, qualitative MOVB theory as formulated in this work is the non-numerical analogue of "state of the art" numerical MO or VB theory.*

It is essential that one understands what the adjectives "conceptual" and "formal" imply:

1. By conceptual advance we mean that MOVB theory can reveal the origin of a chemical trend in a more direct and unambiguous way than SD MO, leave alone polydeterminantal MO and VB, theory. This is due to the fact that each MOVB CW has chemical significance, each matrix element can be decomposed into the chemically meaningful terms F or P, G, and X, each MOVB state can be conveniently expressed by bond diagram(s), and all arguments are "total state energy" arguments. Luxuries of this type are not afforded by SD MO theory.

2. By formal advance we mean that MOVB theory can reveal the origin of a chemical trend while SD MO theory cannot do so because of the approximations made at this level of theory.

It follows that, by applying qualitative MOVB theory to a variety of problems, one has the opportunity to identify conceptual as well as formal breakdowns of qualitative SD MO theory. Thus, in some cases, we will find that although SD MO theory correctly computes A to be more stable than B, the accompanying interpretation is incorrect. In other cases, we will find that conclusions to which one is led by application of SD MO theory are actually meaningless as the electronic effect truly responsible for the chemical trend in question is not "contained" within SD MO theory. Finally, there will be cases in which MOVB theory has a conceptual and a formal advantage over SD MO theory. Inthis work, we have provided illustrative applications of MOVB theory to problems which, in principle, can be satisfactorily dealt with by SD MO theory, i.e., we have sought to convince the reader of the conceptual superiority of MOVB theory. In following papers, we shall exploit the formal as well as conceptual advantages of MOVB theory.

L. Epilogue

There is a certain inevitability of advancement in science. Specifically, scientific evolution can be viewed as a constant alternation of two phases of human endeavor. In the first, passive, phase, one is content with the exploration of existing models and ideas. In the second, active, phase, an accumulation of difficulties besieging the existing frameworks motivates a revolution. Usually, this is initiated in an unobtrusive manner and it is instigated by the work of more than one individual. Once the revolution has run its course one invariably wonders as to why it did not occur earlier! That is to say, we recognize, in an a posteriori sense, the inevitability of the institutional change itself. It is not difficult to identify developments of this magnitude in chemistry. For example, a transition from a two dimensional to a three dimensional view of molecules signaled the advent of conformational analysis. A preoccupation with straight chain polymers was replaced by a more general stereochemical view of polymer structure, which gave due consideration to helical symmetry. This led to the elucidation of the structure of nucleic acids and proteins. The concept of the two electron-two center bond was replaced by the broader viewpoint of many electron-many center bonds. The chemical implications of the Hückel MO theory were recognized many years after the publication of the original theory, and so on. In this paper, we have asked the obvious question: Can we construct a comprehensive qualitative theory of the chemical bond by removing the key assumptions which have dominated our thinking for decades? We have taken this inevitable step and we have found that the emerging conceptual network does not defy comprehension. Indeed, we can say that we have finally accomplished our goal of formulating a "back of the envelope" rigorous qualitative theory of chemical bonding since all steps which have led us to this point are retraceable to the fundamental equations of VB theory in a way that all approximations are mathematically defined. In other words, we have a refutable qualitative theory of chemical bonding. Henceforth, we shall be using the compact form of qualitative MOVB theory in connection with applications of the theory to diverse chemical problems with the exception of the next paper dealing with the structures of H_2O and H_2O_2 where, for illustrative purposes, we use both the detailed and the compact forms of the theory. We shall see that this type of qualitative theory can be easily applied to large molecules and complexes so that for the first time we can begin to understand bonding of diverse systems (e.g., organic, inorganic, biochemical, etc.) using always the same fundamental concepts. It is in this manner that we hope to annihilate interdisciplinary barriers which now preclude a more catholic view of chemistry and physics.

REFERENCES

1. Parr, R. G. "Quantum Theory of Molecular Electronic Structure", W. A. Benjamin: New York, 1913.
2. The "father" of the FO approximation in qualitative MO theory is K. Fukui. Fukui, K.; Yonezawa, T.; Shingu, H. J. Chem. Phys. 1952, 20, 722. Fukui, K.; Yonezawa, T.; Nagata, C.; Shingu, H. J. Chem. Phys. 1954, 22, 1433.
3. For applications of the FO-Perturbation MO (FO-PMO) model to problems of chemical reactivity, see:
   (a) Dewar, M.J.S. "The Molecular Orbital Theory of Organic Chemistry", McGraw-Hill: New York, 1969.
   (b) Hudson, R.F. Angew. Chemie, Int. Ed. Engl. 1973, 12, 36.
   (c) Klopman, G. in "Chemical Reactivity and Reaction Paths", Klopman, G., Ed.; Wiley-Interscience; New York: 1974.
   (d) Fleming, I. "Frontier Orbitals and Organic Chemical Reactions", John Wiley; New York: 1976.
4. A one-electron MO perturbation theoretical approach to structural chemistry is described in: Epiotis, N.D.; Cherry, W.R.; Shaik, S.; Yates, R.L.; Bernardi F. Top. Curr. Chem. 1977, 70, 1. In this work, frequent use of the FO approximation is made.
5. Fukui, K. "Theory of Orientation and Stereoselection"; Springer-Verlag: Berlin and New York, 1975.
6. (a) Woodward, R.B.; Hoffmann, R. J. Am. Chem. Soc. 1965, 87, 395.
   (b) Longuet-Higgins, H.C.; Abrahamson, E. J. Am. Chem. Soc. 1965, 87, 2045.
   (c) Hoffmann, R.; Woodward, R.B. J. Am. Chem. Soc. 1965, 87, 2046.
   (d) Woodward, R.B.; Hoffmann, R. "The Conservation of Orbital Symmetry"; Verlag Chemie: Weinheim, 1970.

7. (a) Dewar, M.J.S. Angew. Chem., Int. Ed. Engl. 1971, 10, 761.

   (b) Zimmerman, H.E. Acc. Chem. Res. 1971, 4, 272.

8. The independent recognition of orbital symmetry control of reaction stereochemistry by a number of brilliant investigators is documented in a recent article: Epiotis, N.D.; Shaik, S.; Zander, W. in "Rearrangements in Ground and Excited States", Vol. 2, de Mayo, P., Ed., Academic Press, Inc.: New York, 1980.

9. A definition of "legitimate" and "illegitimate" exceptions of rules is given in ref. 1.

10. The "father" of the concept of aromaticity is E. Hückel: Hückel, E. Z. Physik, 1931, 70,204; ibid., 1932, 76,628. Hückel, E. Z. Electrohem. 1937, 43,752. Its applicability to problems of chemical reactivity and, in particular, pericyclic reactions was recognized independently under different theoretical disguises by M. G. Evans, M. J. S. Dewar, and H. E. Zimmerman.

    (a) Evans, M.G. Trans. Faraday Soc. 1939, 35, 824.

    (b) Ref. 7

11. Pauli, W. Z. Physik 1925, 31, 765.

12. (a) Mulliken, R.S. J. Am. Chem. Soc. 1950, 72,600; 1952, 74,811; J. Phys. Chem. 1952, 56, 801.

    (b) Mulliken, R.S.; Person, W.B. "Molecular Complexes"; Wiley-Interscience: New York, 1969.

    See also: Murrell, J. N. Quart. Reviews 1961, 15, 191.

13. (a) Förster, Th. Angew. Chem., Int. Ed. Eng. 1969, 8, 333.

    (b) Beem. J.; Knibbe, H.; Weller, A. J. Chem. Phys. 1967, 47, 1183.

    (c) "The Exciplex", Gordon, M.; Ware, W. R., Eds; Academic Press, Inc: New York, 1975.

14. (a) Mason, R. Nature 1968, 217, 543.

(b) McWeeny, R.; Mason, R.; Towl, A. D. C. Discuss. Faraday Soc. 1969, 47, 20.

(c) Mason, R. IUPAC Int. Congr. Pure Appl. Chem., 13th 1971, 6, 31.

(d) Drago, R. S.; Corden, B. B. Acc. Chem. Res. 1980, 13, 353.

15. Fukui, K.; Fujimoto, H. Bull. Chem. Soc. Japan, 1968, 41,1989; 1969, 42, 3399.

16. Epiotis, N.D. Angew. Chemie, Int. Ed. Engl. 1974, 13, 751.

17. (a) Epiotis, N.D.; Shaik, S. Prog. Theor. Org. Chem. 1977, 2, 348.

(b) Epiotis, N.D.; Shaik, S. J. Am. Chem. Soc. 1978, 100, 1 and subsequent papers.

(c) Epiotis, N.D., "Theory of Organic Reactions"; Springer-Verlag: Berlin and New York, 1978.

18. (a) Moffitt, W. Proc. Roy. Soc. 1951, A210, 224, 25.

(b) Balint-Kurti, G.G.; Karplus, M. J. Chem. Phys. 1969, 50,478.

19. (a) Glasstone, S.; Laidler, K.; Eyring, H. "The Theory of Rate Processes"; McGraw-Hill: New York, 1941.

(b) Ellison, F.O. J. Am. Chem. Soc, 1963, 85,3570; J. Chem. Phys. 1964, 41,2198.

(c) Blais, N.C.; Truhlar, D.G. J. Chem. Phys. 1973, 58,1090.

(d) Steiner, E.; Certain, P.; Kwitz, P. J. Chem. Phys. 1973, 59,47.

(e) Tully, J.C. J. Chem. Phys. 1973, 58,1396; 1973, 59, 5122.

(f) Cashion, J.K.; Herschbach, D.R. J. Chem. Phys. 1964, 40,2358; 1964, 41, 2190.

20. For review of "Molecules in Molecules" theoretical approaches, see: Fabian, J. J. Signal AM 6 1978, 4, 307; Fabian, J. J. Signal AM 7 1979, 1, 67.

21. For example, see: Murrell, J.N. in "Orbital Theories of Molecules and Solids", March, N.H., Ed.; Clarendon Press: Oxford, 1974.

22. The explicit forms of the X_a's can be found in texts of quantum chemistry. See, inter alia: Richards, W.G.; Horsley, J.A. "Ab Initio Molecular Orbital Calculations for Chemists"; Clarendon Press: Oxford, 1970. It is important to keep in mind that the selection of the linearly independent VB-type wavefunctions is arbitrary when discussing the physical significance of VB matrix elements.

23. The total Hamiltonian for a composite system made up of fragments C and L can be written as $\hat{H}_C+\hat{H}_L+\hat{I}$, where $\hat{H}_C$ acts on fragment C, $\hat{H}_L$ on fragment L, and $\hat{I}$ is the interaction operator. The MO's of fragment C are orthogonal because they are eigenfunctions of $\hat{H}_C$ and the same is true for the MO's of fragment L which are eigenfunctions of $\hat{H}_L$.

24. Wolfsberg, M.; Helmholz, L. J. Chem. Phys. 1952, 20, 837.

25. For review of approximations of β_{tu} resonance integrals and lucid discussions of the HMO and EHMO methods, see: McGlynn, S.P.; Vanquickenborne, L.G.; Kinoshita, M.; Carroll, D.G. "Introduction to Applied Quantum Chemistry"; Holt, Rinehart, and Winston, Inc.: New York, 1972.

26. In practice, there is only magnitude dependence in the case of cyclic AO overlap. This point can be understood by rewriting H_{ii} as:
$H_{ii} = (F_i + G_i) + [f(s^{even}) + f(s^{odd})]$
where F_i is the energy of the isolated fragments. The term in parenthesis is always larger, in absolute magnitude, than the term in brackets. Thus, the sign and magnitude fluctuations of the latter cannot change the overall sign of H_{ii}.

27. The MOVB theory of diradicals is developed in: Epiotis, N.D. Pure and Appl. Chem. 1979, 51, 203.

28. Pauling, L. "The Nature of the Chemical Bond"; Cornell University Press: Ithaca, New York, 1969.

29. The reader will recall that, in an analogous sense, "change transfer", i.e., "interbond ionicity", was found to be a critical differentiating element in the VB treatment of "aromaticity and antiaromaticity" in four electron-four center systems in Part 1.

30. The two fragment MOVB theory of stereoselection is described in references 15-17. The equivalent three-fragment analysis can be found in: Inagaki, S.; Fujimoto, H.; Fukui, K. J. Am. Chem. Soc. 1976, 98, 4693.

31. (a) Hoffmann, R.; Lipscomb, W.N. J. Chem. Phys. 1962, 36,2189.
(b) Hoffmann, R. J. Chem. Phys. 1963, 39, 1397.

32. (a) Apeloig, Y.; Schleyer, P.v.R.; Binkley, J.S.; Pople, J.A. J. Am. Chem. Soc. 1976, 98, 4332.
(b) Collins, J.B.; Dill, J.D.; Jemmis, E.D.; Apeloig, Y.; Schleyer, P.v.R.; Seeger, R.; Pople, J.A. ibid. 1976, 98, 5419.

33. Gole, J.L.; Siu, A.K.Q.; Hayes, E.F. J. Chem. Phys. 1973, 58, 857.

34. Herzberg, G. "Electronic Spectra of Polyatomic Molecules"; Van Nostrand: Princeton, 1966, pp. 491-493.

35. This important trend is due to the fact that the precursor of ω_2 is the antibonding combinations of s_1 and s_2 while the precursor of ω_4 is the bonding combination of p_1 and p_2 where the subscripts refer to the carbon atomic centers. As a result, the energy gap separating ω_2 and ω_4 is fully expected to be much smaller than the energy gap separating the valance s and p AO's of carbon atom.

36. For computations of linear and bent forms of C_2H_2 and the related C_2H_4, see:
a) Strozier, R.W.; Caramella, P.; Houk, K.N. J. Am. Chem. Soc. 1979, 101, 1340.
b) Volland, W.V.; Davidson, E.R.; Borden, W.T. J. Am. Chem. Soc. 1979, 101, 533.

Postscript

It is the predicament of an author who writes based on his own set of preconceptions that he makes occasional statements which could be potentially misunderstood by the readership. In response to the very enlightened comments of one of the editors, we would like to reemphasize one important thing: This work is not a polemic against MO theory! If one desires to understand chemistry in terms of a rigorous mathematical model and obtain numerical results, MO theory has served us and continues to serve us in a marvelous way. This work offers no arguments or opinions as to what is the best way of numerically computing atomic and molecular properties. On the other hand, if one desires to achieve a qualitative understanding of chemistry in a "back of the envelope" sense, MO-CI theory becomes conceptually too cumbersome. As we point out in the introductory part, this problem of conceptualization persists at lower levels of MO theory, even at the level of EHMO theory. Thus, when we say that our previous "understanding" of chemistry has been often "illusory", we refer to some (not all) Hückel-type and one-electron perturbation MO treatments based on the approximations which we have now rejected: Frontier Orbital approximation, low order Perturbation Theory, etc. Again, we must be careful: This last statement should not be taken as a condemnation of Hückel-type theory. For there are problems for which this brand of theory is perfectly suitable and there are practitioners who are elegantly using these methods with a good understanding of their limitations. Our own goal has been to introduce a compromise MOVB language which, in fact, has led us to revise most of our thinking as to the "how's and why's" of chemistry. This will become much clearer as a following series of papers unfolds.

Aside from the danger of having his intentions misunderstood, there is a second inevitable problem which is associated with any effort to present a new approach. This has to do with the proper recognition of related contributions of theoretical chemists which have not been very visible in the arena of applied theory. This author is keenly aware of them but also is the victim of space limitations, in addition to being starkly aware of the dislike of experimentalists for equations and formalism, in general. In this light, important works by Ruedenberg[a], Kutzelnigg[b], Kollmar[c], Freed[d], and others dealing with total energy decomposition schemes,as well as important papers on VB formalism and applications[e], have not been discussed because this would create an additional barrier for the nonexpert to overcome on his way to the MOVB theory and its applications, the main contribution of this work.

In conclusion, a manuscript of this type is undoubtedly bound to offend the sensibilities of some. For after all, the problem of chemical bonding is an ancient problem which has occupied the attention of some of the best minds of natural science. As a result, many credible views have been expressed in the literature. Ours is simply one more view which has crystallized after considerable entanglement with chemical structure and reactivity problems.

a) Ruedenberg, K. Rev. Mod. Phys. 1962, 39, 326.

b) Driessler, F.; Kutzelnigg, W. Theor. Chim. Acta 1967, 43, 1, 307. Kutzelnigg, W."Einführung in die Theoretische Chemie, B. II. Die Chemische Bindung"; Verlag Chemie: Weinheim, 1978.

c) Kollmar, H. J. Am. Chem. Soc. 1978, 101, 4832; Kollmar, H. Theor. Chim. Acta 1978, 50, 235; ibid., 1980, 58, 19.

d) Freed, K. in "Modern Theoretical Chemistry", Vol. 7, G. A. Segal, Ed., Plenum: New York, 1977, p 201.

e) For bibliography and excellent review of explicit and implicit VB methods, see: Pauncz, R. "Spin Eigenfunctions"; Plenum Press: New York, 1979.

A. F. Williams

A Theoretical Approach to Inorganic Chemistry

1979. 144 figures, 17 tables. XII, 316 pages
ISBN 3-540-09073-8

Contents: Quantum Mechanics and Atomic Theory. – Simple Molecular Orbital Theory. – Structural Applications of Molecular Orbital Theory. – Electronic Spectra and Magnetic Properties of Inorganic Compounds. – Alternative Methods and Concepts. – Mechanism and Reactivity. – Descriptive Chemistry. –Physical and Spectroscopic Methods. – Appendices. – Subject Index.

This book outlines the application of simple quantum mechanics to the study of inorganic chemistry, and shows its potential for systematizing and understanding the structure, physical properties, and reactivities of inorganic compounds. The considerable strides made in inorganic chemistry in recent years necessitate the establishment of a theoretical framework if the student is to acquire a sound knowledge of the subject. A wide range of topics is covered, and the reader is encouraged to look for further extensions of the theories discussed. The book emphasizes the importance of the critical application of theory and, although it is chiefly concerned with molecular orbital theory, other approaches are discussed. This text is intended for students in the latter half of their undergraduate studies. (235 references)

Springer-Verlag
Berlin
Heidelberg
New York

Lecture Notes in Chemistry

Vol. 1: G. H. Wagnière, Introduction to Elementary Molecular Orbital Theory and to Semiempirical Methods. V, 109 pages. 1976.

Vol. 2: E. Clementi, Determination of Liquid Water Structure. VI, 107 pages. 1976.

Vol. 3: S. R. Niketic and K. Rasmussen, The Consistent Force Field. IX, 212 pages. 1977.

Vol. 4: A. Graovac, I. Gutman and N. Trinajstić, Topological Approach to the Chemistry of Conjugated Molecules. IX, 123 pages. 1977.

Vol. 5: R. Carbo and J. M. Riera, A General SCF Theory. XII, 210 pages. 1978.

Vol. 6: I. Hargittai, Sulphone Molecular Structures. VIII, 175 pages. 1978.

Vol. 7: Ion Cyclotron Resonance Spectrometry. Edited by H. Hartmann and K.-P. Wanczek. VI, 326 pages. 1978.

Vol. 8: E. E. Nikitin and L. Zülicke, Selected Topics of the Theory of Chemical Elementary Processes. X, 175 pages. 1978.

Vol. 9: A. Julg, Crystals as Giant Molecules. VII, 135 pages. 1978.

Vol. 10: J. Ulstrup, Charge Transfer Processes in Condensed Media. VII, 419 pages. 1979.

Vol. 11: F. A. Gianturco, The Transfer of Molecular Energies by Collision: Recent Quantum Treatments. VIII, 328 pages. 1979.

Vol. 12: The Permutation Group in Physics and Chemistry. Edited by J. Hinze. VI, 230 pages. 1979.

Vol. 13: G. Del Re et al., Electronic States of Molecules and Atom Clusters. VIII, 177 pages. 1980.

Vol. 14: E. W. Thulstrup, Aspects of the Linear and Magnetic Circular Dichroism of Planar Organic Molecules. VI, 100 pages. 1980.

Vol. 15: A.T. Balaban et al, Steric Fit in Quantitative Structure-Activity Relations. VII, 178 pages. 1980.

Vol. 16: P. Čársky and M. Urban, Ab Initio Calculations. VI, 247 pages. 1980.

Vol. 17: H. G. Hertz, Electrochemistry. X, 254 pages. 1980.

Vol. 18: S. G. Christov, Collision Theory and Statistical Theory of Chemical Reactions. XII, 322 pages. 1980.

Vol. 19: E. Clementi, Computational Aspects for Large Chemical Systems. V, 184 pages. 1980.

Vol. 20: B. Fain, Theory of Rate Processes in Condensed Media. VI, 166 pages. 1980.

Vol. 21: K. Varmuza, Pattern Recognition in Chemistry. XI, 217 pages. 1980.

Vol. 22: The Unitary Group for the Evaluation of Electronic Energy Matrix Elements. Edited by J. Hinze. VI, 371 pages. 1981

Vol. 23: D. Britz, Digital Simulation in Electrochemistry. X, 120 pages. 1981.

Vol. 24: H. Primas, Chemistry, Quantum Mechanics and Reductionism. XII, 451 pages. 1981.

Vol. 25: G. P. Arrighini, Intermolecular Forces and Their Evaluation by Perturbation Theory. IX, 243 pages. 1981.

Vol. 26: S. Califano, V. Schettino and N. Neto, Lattice Dynamics of Molecular Crystals. VI, 309 pages. 1981.

Vol. 27: W. Bruns, I. Motoc, and K. F. O'Driscoll, Monte Carlo Applications in Polymer Science. V, 179 pages. 1982.

Vol. 28: G. S. Ezra, Symmetry Properties of Molecules. VIII, 202 pages. 1982.

Vol. 29: N.D. Epiotis, Unified Valence Bond Theory of Electronic Structure. VIII, 305 pages. 1982.